Networking Peripheries

Networking Peripheries

Technological Futures and the Myth of Digital Universalism

Anita Say Chan

The MIT Press
Cambridge, Massachusetts
London, England

This book was set in Stone Sans and Stone Serif by the MIT Press.

Library of Congress Cataloging-in-Publication Data

Chan, Anita.
Networking peripheries : technological futures and the myth of digital universalism / Anita Say Chan.
  pages   cm
Includes bibliographical references and index.
ISBN 978-0-262-01971-2 (hardcover : alk. paper), 978-0-262-55207-3 (paperback)
1. Information society—Peru. 2. Information technology—Peru. 3. Digital divide—Peru. 4. Technological innovations—Social aspects—Peru. I. Title.
HN350.Z9I5634   2013
303.48'330985—dc23
2013006956

Dedicated to Arturo and Rosario Say

Contents

Preface

The question has been posed to me more than a few times: why study digital culture and information technology (IT) in Peru? Within global imaginaries, Peru evokes the mountainous South American nation that once served as the heart of the Inca civilization: home to Machu Picchu, high stretches of the Andes, and large populations of Quechua- and Aymara-speaking communities. Or, as the source of raw, natural commodities, such as the world's largest cultivations of coca leaves—from which cocaine is infamously drawn—and some of the largest reserves of gold, silver, and copper, whose promise first lured Spanish conquistadores to South America in the early sixteenth century. This Peru might be known as an ideal space from which to peer into past tradition, native culture, or the plethora of nature's bounty - but what could it tell us about the dynamics of contemporary digital culture, high technological flows, or their associated future-oriented developments?

The question itself, of course, merits unpacking, given it often assumes that digital culture—despite its uniquely global dimensions—does indeed have more legitimate, productive sites from which to undertake its study. Foremost among them are the labs, offices, and research sites nestled in Silicon Valley and their dispersed equivalents in other innovation capitals worldwide. It is there that digital culture presumably originates and has its purest form—only to be replicated elsewhere; there that visions for digital futurity in their most accurate or ideal approximations emerge; and there that technological advancements—and thus digital cultural advancements—are understood to be at their most dynamic, lively, and inspired. Although places such as Peru unlock a path strewn with the relics and treasures of a technological past one must struggle not to forget, sites such as Silicon Valley unlock the secrets to a technological future whose path has yet to be fully tread, even though it has already become the source of fevered cultural obsession.

Rarely does anyone have to justify choosing such exclusive places to represent "digital culture" writ large. Just as rarely is there a need to explain the implication that the digital futures imagined by select populations of engineers, designers, and innovators of new technologies in elite design centers can—or should—speak for the global rest, and that the present currently unfolding in innovation centers must surely be the future of the periphery. There is a particular notion of the periphery conjured here, of course, as mere agents of global counterfeit—as sites of replication of a future invented prior and elsewhere. As much as the authoritative role of innovation centers for extending design and invention is rarely questioned, so, too, is the periphery viewed unquestioningly as a zone of diffusion and simple uptake of such designs. But the periphery is hardly so passive or uninventive. Outdoor markets of used, recycled, and reassembled computer parts take up entire streets, Internet cafés pepper neighborhoods with low-cost and ready Internet access, and gray markets of digitized local music and pirated films fill multistory buildings. Such technological hacks and local improvisations are an everyday part of the periphery's technology landscape (Alarcon 2008; Durant 2009; Liang 2003a, 2003b, 2004 2005; Liang and Prabhala 2007; Sundaram 1999) whose vibrancy is only partly captured by comparing them to formal or established commercial markets of digital goods. Just as accounts of the innovation culture of Silicon Valley based only on the research activities, entrepreneurship, and hacks of engineers and IT-related professionals—rich as such accounts have been (Anderson 2006, 2010; Cringley 1996; Hafner and Lyon 1996; Hiltzik 2000; Kidder 2000; Levy 1984, 2000, 2011; Lewis 2001; Markoff 2006; Saxenian 1996; Turner 2006; Vaidhyanathan 2011)—tell only part of that story. By applying methods and techniques that previously have been unapplied—and often unimagined—in innovation centers, these places on the "digital periphery" build new structures and spaces that—for those living in the Global South today—enable the primary means of accessing and interacting with digital technologies. As the social scientists Daniel Miller and Don Slater (2001) observed in their study of Trinidad, "the Internet is not a monolithic cyberspace," but exists instead as a globally expansive technology with various local realities, adoptive practices, and cultural politics that surround its varying localizations. There have been, indeed, more ways than one to imagine what digital practice and connection could look like.

Conserving centers as exclusive sites of origination and invention, of course, also neglects another crucial detail—that the centers of the present were once on the periphery, too. To focus on centers as inventing models that simply come to be adopted and copied elsewhere presumes the perfect, continual extension of replicative functions and forces. It fails to account for the possibility of change within the larger system—the destabilizations and realignments of prior centers—and so, too, the realignments of prior peripheries. The global financial crisis of 2008 and its fallout thereafter provided the latest reminders that the stability of established powers and the permanence of center-periphery relation can be put in question. At a time, indeed, when the power of existing global centers and circumstances seems no longer to be a given, when the invocations of "freefall" (Stiglitz 2010) and the "end of times" (Žižek 2011) have been used by liberal economists and progressive philosophers alike to characterize the contemporary, and when unexpected interruptions and shifts in world dynamics seem to be the only reliable constant, heightened scholarly attention to the creative engagements in digital culture at the periphery, more than ever, seems to be warranted. Far from merely lagging behind or mimicking centers, the dynamic activity from peripheral sites suggests how agents once holding minor status—and even the notion of "the copy" itself—can emerge instead as fresh sources of distinctly optimized or unencumbered productivity.

What follows here, then, aims to trace the vibrant micro narratives and situated stories around IT and innovation cultures multiplying at the periphery. Such diverse threads unsettle the unspoken presumption that a single, universal narrative could adequately represent the distinct digital futures and imaginaries emerging from local sites today. This work explores the extension of such micro narratives, then, not in the sense of being minor in either scale or ambition but rather for departing from conventional frameworks that presume digital imaginaries and futures as necessarily—or even best—represented by innovation centers. Minor narratives, similar to minor chords in musical scales, can be read as complementing and decentering the movements of their major twin.

More than a decade into the new millennium, there is, indeed, little shortage of new social experiments and global collaborations emerging around digital culture from technology's so-called periphery. Although audiences around the world were captivated by the wave of social protests in the Middle East starting in early 2011 and the online corporate protests of the Anonymous global hacker network (Coleman 2011, 2013), popular uses of digital technologies have been associated with social movements since the early years of the twenty-first century and in sites as diverse as Mexico (Schultz 2007), Iran (Burns and Eltham 2009; Grossman 2009), the Philippines (Uy-Tioco 2003; Vincente 2003), and Ukraine (Morozov 2010b). In the wake of other networked social movements that followed the 2011 Arab Spring—from Spain's 15-M Indignados movement to North

American Occupy movements—social networking technologies originally developed and launched years ago by US IT firms seem to have come of age in the hands of users in remote and distant elsewheres.

Media channels worldwide also have registered a new urgency to find explanations for the rapid, globally mimetic spread of such popular uprisings. In the midst of the 2011 Middle East uprisings, Silicon Valley entrepreneurs and designers of the digital applications used by networked protestors were sought out in news reports for their analysis of the growing transnational activity. Yet for all their familiarity with the technologies, they seemed little better prepared for nuanced explanation of the emerging global forms and calls for political change than other commentators on global communication policy—neither of whom seemed to have been asked before to directly address what innovation cultures and digital practices could mean outside of technological centers. Prior explanations recirculated of the Internet's global diffusion as either—by nature—fostering greater liberty and democracy through the extension of technological openness, or fostering precisely its opposite through the extension and acceleration of open markets, economic globalization, and, ultimately, a digitally paced capitalism that served only a limited collective of technological elites in such centers (Anderson 2006, 2010; Castells 1996; Castells, Fernandez-Ardevol, Qiu, and Sey 2009; Kelly 2010; Ross 2004, 2010; Shirky 2009). Both versions, relying on visions of networked technologies as simply diffusing from innovation centers in the West to the rest (Callon and Latour 1981; Latour 1987, 2005) seemed at once entirely familiar and all too partial.

Indeed, this study was partially inspired by the clear diversity of accounts of worldwide digital cultures. Predictions that digital technology would produce either freedom or oppression as defined from centers simply didn't match the contours of digital culture at the periphery. In Peru, such evidence had visibly accumulated by the early 2000s—bringing a range of distinct actors and interests into unexpected and often contradictory proximity. Collectives of free software advocates (who had helped to bring the first UN-sponsored conference on free software use in Latin America—a landmark event—to the ancient Incan capital of Cuzco in 2003) sought to reframe the adoption of open technologies as not just an issue of individual liberty and free choice, as it had been for free, libre, open-source software (FLOSS) advocacy in the United States, but of cultural diversity, state transparency, and political sovereignty from the monopolistic power of transnational corporations in the Global South. "Digital innovation" classrooms installed in rural schools by the state would later be converted into the largest deployment network for MIT's high-profile One Laptop per Child (OLPC) initiative just several years later, all in the name of enabling universal digital inclusion. And intellectual property (IP) titles newly and aggressively applied by state programs on "traditional" goods promised to convert rural producers and artisans into new classes of export-ready "information workers" as part of the nation's growing information society–based initiatives.

Accounts of such engagements would eventually make it into the chapters of this volume; but when I first encountered them as a graduate student years ago, there were few resources in the field of digital or new media studies that could make the range of their investments legible. FLOSS advocates and high-tech activists in Cuzco, state-promoted "innovation classrooms" in rural schools, and traditional artisans as new global "information workers" were not the conventional interests or protagonists that emerged from most of the existing tales spun through centers of digital culture. To watch these "other" stories unfold was to watch the details of each spill over the edges of the existing frameworks and dominant narratives of digital culture. Global imaginaries around IT, now a decade into the new millennium, might have made Silicon Valley hackers, the obsessions and tireless work ethic of high-tech engineers, and strategic enterprise of competitive technology entrepreneurs and innovators the stuff of popular Hollywood films. But to capture the dynamic engagements and fraught experiments in digital culture in Peru still required attention to a host of other stakes, agents, and developments than the increasingly recognizable cast of high-tech heroes and villains allowed. In working around the digital in Peru, such actors inevitably confronted the contradictions of what it meant to build new links and exchanges between spaces of the rural and the urban, the high-tech and traditional, and pitched appetites for the global with intensive commitments to the local.

They were a reminder that to study the Internet across global sites required more than continued attention to the line between online and offline worlds, or to unpacking the development of networked infrastructure and their associated digital technologies, or to the workings of Western-centered engineers and innovators. Valuable as such research trajectories have been, the stakes being carved out across globally-dispersed local sites continue to expand. Information technologies at the beginning of the new millennium, after all, have come to operate as shorthand for imagining global connection in a variety of modern forms—economic, political, scientific, cultural, and, indeed, digital. And in twenty-first century Peru, they have been as much about imaging—and reimaging—relations with the local, rural, and peripheral as they have been about drawing out connections to established cultural and technological centers. Such local and peripheral encounters, as much as those of global connection, prove indeed to be ever-critical stakes of digital engagement that depend on the cultivation of interfaces among distinct interests and communities across spatial, political, or economic divides—and that prove to be anything but merely innocent undertakings.

Cultural ambivalence and technological complexity, much as they do in technological centers, weave diverse and surprising tales around digital connection at the periphery. As I came to see, they involved the interruption as well as expression of new extensions of power. Why they might have escaped attention for so long is part of their story, too.

Global Integrations

In twenty-first-century Peru, there have been no shortage of narratives of global connection and related accounts of new, radical transformations.

Since the new millennium, Peru has ridden a wave of new economic growth that reversed the economic stagnation it saw during much of the previous century, notably defying the 2008 financial crisis that spread across Europe and North America. World Bank officials credited such performance to Peru's pointed expansion of free trade agreements—more than two dozen to date—with partners including the United States, Chile (both signed in 2006), Canada, China, Japan (signed in 2009, 2010, and 2011, respectively), and South Korea (2011). Such moves to increase global economic participation are what reportedly allowed it to establish a new record as "a top performer in Latin America"—and one of the few nations "in the region to record positive growth (0.9%) [during a period that saw stagnation as the primary effect of the global financial crisis], performing better than the average for Latin America (-2.3%)" (World Bank 2012).

The International Monetary Fund (IMF) reports similarly celebrated Peru's economic reemergence. After decades of military and authoritarian rule, a divisive armed conflict in the 1980s and 1990s with the radical Maoist peasant movement Sendero Luminoso that turned the countryside—and countless villages with it—into war zones of the most extreme political violence the nation had seen since it won independence from Spain, the new millennium seemed to have at last ushered in a new beginning and turn toward renewal. The IMF reported that Peru's economy in 2007 experienced the largest growth rate, 9 percent, of any Latin American nation, a result repeated in 2008 with growth measured at 9.8 percent. IMF records released at the end of 2010 and 2011 similarly reported growth at 8.75

percent and 6.75 percent, respectively—making it a leader on the continent, where growth averaged 4.7 percent in 2011, and marking it as one of the world's fasting growing economies in the last decade. The IMF's positive outlook continued with 2012 estimates of 5.25 percent growth (International Monetary Fund 2011; Vigo 2011).

The recent economic upswing has translated into the expansion of newly active and visible consumer classes in Lima and other urban centers. Gleaming imported sedans and a record-breaking construction of apartment buildings and new shopping centers peppered with global brands from Sony to Starbucks can be seen in middle-class neighborhoods and Lima's working-class peripheries alike. Imported consumer electronics likewise are becoming more and more commonplace across Peru's outer provinces. If divides of cultural, economic, and regional difference still undeniably mark much of contemporary Peruvian society, an appetite for new consumer and electronic goods in Lima seems to be at least one realm where such differences have had the chance to suddenly dissolve.

Mobile information technologies boast new visibility, too. Although roughly a third of the residents in Peru's capital city of Lima had cell phones at the end of 2004, by 2010, some 83.3 percent did. In urban areas outside Lima for the same period, mobile phone use jumped from 14.7 percent to 81.3 percent. And in rural areas, it increased from just 1.3 percent to 46.2 percent. By 2011, some 74.3 percent of households nationwide were estimated to have a cell phone according to Peru's National Institute of Statistics (INEI). The expansion of digital connectivity is even more noticeable in rural zones when measured against traditional phone lines. In towns with fewer than two thousand inhabitants, landlines were maintained by 2.7 percent of households, whereas some 46.9 percent had cell phones. Even if it is still far from all inclusive, digital culture in new millennial Peru could hardly be ascribed only to an established elite—at least not in its former, rurally excluding guise.

The policies to which much of the recent growth has been attributed, in fact, had been put in place more than a decade earlier under Alberto Fujimori's presidency (1990–2001). It was then that a wide range of neoliberal policies were enacted—from the privatization of public enterprises, to the encouragement of foreign investment, to the selling of public lands to mining and energy extraction sectors. Despite being democratically elected, Fujimori would later be remembered for his authoritarian autogolpe (self-coup) that suspended constitutional rule, shut down Congress, and purged the judiciary. He would also be remembered for his record of human rights violations, political disappearances, and kidnappings during the civil war against Sendero Luminoso (during which, some sixty-nine thousand Peruvians are known to have been killed), and for routine practices of embezzlement and corruption that would later lead to his impeachment, escape to Japan in 2001, and eventual extradition and imprisonment in Peru in 2007.

The free market policies he put in place, however, like the man himself, still found considerable national support, especially within urban entrepreneurial sectors. Notably, the following presidential administrations of Alejandro Toledo (2001–2006) and Alan Garcia (2006–2011) maintained and arguably even accelerated Fujimori's free market policies. Despite Toledo's and Garcia's public criticisms of Fujimori's political record, his economic policies, particularly when global growth was involved, resisted critique.

Change in Peru's urban zones from such neoliberal economic policies was only beginning to be visible in 2003, the year I arrived to begin graduate fieldwork. I came to attend the first Latin American and Caribbean Conference on Free Software Development and Use (LACFREE), a conference organized in partnership with the United Nations by an international committee of representatives from Latin America's growing base of FLOSS advocates. As one of the first events to be hosted on the continent addressing regional FLOSS use and development, the conference drew a diverse body of attendees from South and North America and Europe—including university researchers, state officials, NGO and corporate IT representatives, and participants in FLOSS associations such as the Peru Linux User's Group (PLUG) and Computer Professionals for Social Responsibility's (CPSR) Peru branch. Held in Cuzco on the seventeenth-century campus of the National University of St. Anthony Abad of Cuzco and co-organized by several state offices—from the National Advisory on Science, Technology, and Technological Innovation (CONCYTEC) to the Ministry of Education and INEI— the conference kept the visiting information classes abuzz with nearly a week's worth of activities in the ancient Incan capital.

Large groups of college-age FLOSS activists from Peru's provinces and other Latin America countries who had traveled for days by bus—the most affordable but undeniably time-consuming means of travel across South America—stretched out across the conference hall. They had the look of tech-savvy backpackers, with laptops perched on unwashed jeans, backpacks slung under seats, and mountain hiking boots loosely laced around their ankles. Between conference talks, they exchanged tips on where to find low-cost hostels, which bus lines provided the fastest and safest travel between national borders and across mountain routes, and updates on free software coding projects cropping up on campuses across the continent. These were not the typical topics of conversation I was accustomed to hearing among the IT professionals, entrepreneurs, and tech journalists who made up large portions of the attendees of FLOSS events in the United States. There, conferences during the past several years had grown formalized enough to attract the sponsorship of IBM, Red Hat, and other enterprises supporting FLOSS development. Admission fees upward of several hundred dollars were becoming increasingly common, making more and more FLOSS events restricted to smaller segments of information professionals—and generally excluding those likely to travel by bus.

Similar to US FLOSS conferences, however, many of the Cuzco conference's attendees came to see—free of charge—the global FLOSS celebrities who came from the United States to participate. These included Bruce Perens, John "Maddog" Hall, and Richard Stallman, known as the father of the FLOSS movement for his contributions to the GNU platform in the 1980s. Many, however, also came as part of a regional activism that had been spurred around Edgar Villanueva—a senator from the Andean town, Andahuaylas. His sponsorship of one of Latin America's first FLOSS bills and lucid, twelve-page letter written to Microsoft to critique corporate IT monopolies in the Global South made him a hero of sorts among FLOSS advocates in South America and—following the letter's wide circulation in online news channels—places beyond. It was partly due to Villanueva and the FLOSS bill that Peru was chosen to host the first LACFREE event, despite highly visible FLOSS projects in other Latin American nations.

Indeed, Villanueva's bill and his subsequent exchange of letters with Microsoft representatives in Peru forcefully laid out a number of arguments explaining the unique interests citizens of the global periphery had in FLOSS. Such concerns were also expressed in the official 2003 Cuzco statement issued by conference participants, which underscored themes of social inclusion, ethnic and linguistic diversity, and mutual aid. It opened by specifying what were framed as FLOSS's greatest benefits, including "that the freedoms extended to the user of free software enable the possibility of opting out of the simple role of a technological consumer, to become an active participant in the knowledge society; that the license policies of proprietary software aren't sustainable for economies of developing nations; that free software respects the necessity to preserve multi-lingualism and diverse cultural identities in cyberspace," and—perhaps above all—"that free software is an integral part of the construction of a free, just, ethical, and inclusive society in which people have the possibility of mutual and collective aid." (LACFREE 2003) It closed by arguing for the construction of spaces and tools for civic participation and political transparency in issues of state decision making, particularly—but not only—when having to do with technology.

That same year, the values of government accountability and political transparency that FLOSS activists advanced found another strong platform in national debates. Late in 2003, Peru's newly convened Truth and Reconciliation Commission released the official death count from the country's conflict with Sendero Luminoso. Rural populations were found to have suffered most from the violence, which occurred at the hands of state security forces and guerillas alike. And apart from this, there remained persistent public doubts about government accountability, social inclusion, and transparency concerning the "official account" of the war. Although suspected by many to be a conservative estimate, the official release in 2003 of the sixty-nine-thousand death count from the conflict hoped to finally resolve this chapter of the state's long struggle with peasant groups. Meanwhile, over the next decade, various projects stressing the use of information technologies as a means of cultivating a new and inclusive information society proliferated across the nation—with their target, after decades of regional conflict, ambitiously set on national reform, peace, and integration.

The eclectic mix of actors I encountered inside the Cuzco auditorium in 2003 notably reflected the diverse interests I continued to find working in IT in Peru in the years that followed. Drawing together student activists, hobbyist programmers, NGOs, government workers, provincially based teachers, rural producers and communities, representatives from transnational IT companies, and young entrepreneurs, such IT-centered projects were the focus of my research from 2003 to 2011. During that time, I conducted dozens of interviews—more than one hundred in all—with a range of such actors, joining them at their places of work, planning, and coordination and at public presentations or conferences—frequently traveling with them to rural sites of deployment as they worked to develop new IT initiatives and projects of political and social reform. Their accounts weave through the body of this work.

A number of their activities involved engagement with new state-based initiatives to integrate diverse populations into twenty-first-century circuits of exchange. Such was the case with the new Commission for the Development of Information Society—or CODESI—that the state convened in 2005. The Ministry of Tourism and Trade's (MINCETUR) and the National Institute for the Defense of Competition and Protection of Intellectual Property's (INDECOPI) recently launched initiatives to develop new IP titles for rural producers and artisans—the focus of the first section of this book—likewise echoed such hopes. Much as new state investments around digital education that culminated in the plan to deploy MIT's OLPC in all of Peru's primary schools, did too.

Other IT-centered initiatives, however, were citizen launched. Advocacy for Peru's FLOSS bill, similar to the transnational organizing involved in drafting it, manifested emergent networked citizen coordinations. So, too, were later mobilizations around the OLPC's XO laptops that involved organizing with hundreds of teachers across rural provinces in efforts to localize technological deployments and pedagogical orientations in schools. Such novel citizen-centered engagements and spatial coordinations became the focus of chapters in the second section of this book.

While such expressions of digital culture in-the-making have clearly distinct orientations and visions for reform, I bring them together in this volume for a common ambition they shared to shape the nation's informatic future. Whether state or citizen launched, such projects aimed to build new interfaces among diverse sets of multiply situated actors drawn in around the work and projects of technology-based and information-centered reform. Many project participants came on a volunteer basis, and almost all would eventually discover themselves in collaboration with actors they'd rarely if ever expected to work with. Rural teachers alongside transnational digital education advocates, indigenous cultural leaders with open technology activists, and traditional artisans with IP experts and international exporters were drawn into Peru's array of new IT-centered partnerships. And whether state or citizen-centered, such projects necessarily involved the cultivation of new forms of digital and social connection that stretched out toward scales of the global and hyper local sites at once.

That the periphery emerged as a vital stake for new IT investments should perhaps not be surprising. It is at the periphery, after all, where connectivity is commonly projected as finding its greatest challenge. A point not to be taken lightly, I found, given that to follow IT projects in new millennial Peru was to follow their undertakings through zones marked as much by the glowing promise of future reform as they were haunted by the histories of past conflict.

Connecting the Nation

Nearly a decade after Peru's Truth and Reconciliation Commission's 2003 report, Lima's information classes evince an almost palpable optimism about continuing economic growth and global market integration. Across the city, new multistory complexes exclusively dedicated to IT services—and brandishing names such as the Compu-Palace and Compu-Plaza—accommodate growing sectors of information classes. BlackBerries and an array of smartphones, handheld games, and mobile technologies accompany riders during trips on the new eco-ready bus line and elevated electric train stretching across the capital. And the popular new Virtual Museum of Lima, celebrated as one of the world's first museums to be designed as a primarily "virtual" exhibit showcasing digital audiovisual technologies, continues to attract huge crowds, with admission sold out for months after it was first opened in late 2011.

Such recent growth notwithstanding, Peru still had a far longer trajectory of economic, cultural, and geographic divides to overcome. Some 19 percent of the national population was estimated as living in extreme poverty, with an income of less than $2 a day, in 2004 (World Bank 2005). Such rates are significantly higher among rural and indigenous populations—about 29 percent and 32 percent of the nation's twenty-nine million citizens (UNESCO Institute for Statistics 2011), respectively. Although 72 percent of rural populations were measured in poverty and 40 percent in extreme poverty, urban populations registered rates of 40 percent and 8 percent, respectively.

Contemporary Peru, generations after Spanish rule ended in 1821, still bears many of the social, cultural, racial, and geographic stratifications that nearly three hundred years of colonialism wrought. Urban zones still concentrate the majority of the population self-identified as of European descent (15 percent of the national population) or *mestizo*—meaning of mixed European and native descent (45 percent). And the proud identification with European and Western influences often expressed by Lima's elite classes and governing circles starkly contrasts with their ambivalent, conflicted relationship to Peru's mountain and jungle provinces. Of course, Lima, as the first (and once largest) center of governance for the Spanish crown in South America did afford it a certain stature. Originally dubbed "City of Kings" by Francisco Pizarro when it was founded in 1535, it remained a primary center of elite European culture on the continent for centuries until colonial rule was overthrown. Today, however, it's been Peru's rural zones that are recognized to have fueled its contemporary economic growth. Between 1990 and 1997, as investment in mining exploration impressively increased 90 percent worldwide, it grew fourfold in Latin America and twentyfold in Peru alone. By 2003, mining accounted for 57 percent of all exports in Peru. Meanwhile, during the 1990s areas affected by mining concessions increased from four million to sixteen million hectares—so that by decade's end, 55 percent of Peru's six thousand or so *campesino* (peasant) communities were in zones influenced by mining.

And although global integration has been celebrated for generating new economic growth in Peru's urban centers, it has created quite a distinct reaction in the provinces, where social conflicts have risen exponentially since neoliberalization policies were enacted in the 1990s. Over the last decade, the number of protests resulting in deaths of citizens and authorities has mounted, with territorial disputes with rural communities now routine parts of daily news cycles. In the last several years alone, Cajamarca, Piura, and Amazonas have gained notorious reputations as zones of natural extraction and conflict. Each generated heightened public attention after a series of prolonged confrontations escalated to pitched violence. To merely mention Bagua, Conga, or Rio Blanco—the names of extraction projects— summons the memory of protracted conflict for urban and rural citizens alike. More than a decade after conflict with Sendero Luminoso officially ended, discourse of peasant communities opposed to modernization still waxes familiar in the national press—enough so that when rural uprisings do occur, it's all too common to encounter reports that slyly suggest an alliance between rural communities and antistate, terrorist sympathizers; the key message being how rural organizing can subvert Lima's interest in national progress and how distant rural communities remain from circuits of rational discourse and reliable information.

This has not been the only means, however, by which urban information classes have framed relations with their rural other. Cultural divides and conflicts notwithstanding, it is the narrative of national unity, connection, and integration that is underscored in other spaces of new millennial Lima. The Virtual Museum offers one triumphant vision of the city's social cohesion and pluralism, despite the challenging aftermath of colonization and civil conflict, for instance. The museum, built in the historic center of Lima on the site of the former Ministry of Communication, boasts large 3D projections, looming holograms, life-size models, and special effects to turn some thirty rooms into dynamic, "virtualized" spaces to showcase select portions of Lima's ten-thousand-year-long history. Filled with panoramic, wall-to-wall, linked projections that flood rooms with various re-created scenes—from battles between pre-Colombian peoples to Pizarro's sixteenth-century founding of the "City of Kings"—the museum immerses viewers in a spectacle of virtualization to dramatize Lima's evolution.

The exhibits begin with a projection of the marinera dance—the national dance of Peru that originated in coastal creole traditions—on the interior surface of the ministry building's dome arching some four stories overhead. Next is a panoramic, room-encasing projection of Lima's Plaza de Armas. Between shifting, cinematic scenes that stream across the room's long walls, viewers are told to note a modern-day couple whose eyes and smiles first meet across the plaza and whose unfolding romance we're told we will see again in the museum. Meanwhile, further images flash by refer-encing Lima's historic global linkages and cosmopolitan character: brightly colored horse-drawn carriages, the Presidential Palace and cathedral dating back to colonial rule, foreign tourists taking photos, and Lima's governing and working classes mingling in between.

It is oddly less Lima's actual history than the virtual technologies them-selves, though, that the museum arguably best showcases. A cinematically re-created battle scene between the Lima and Wari pre-Colombian peoples, for instance, featuring a human sacrifice ritual and 3D-rendered flying axes and spears, opens with a projected disclaimer that the scene is "not based on any historical records" or official documents but is instead an informed speculation on what may have taken place. Another exhibit re-creates the 1746 earthquake that destroyed large districts of Lima, with a platform shaking beneath visitors' feet as a scene plays depicting the crumbling of the colonial-era court of Lima's vice-regal. No particular details are given of how many citizens died, how far away damages were registered, and who the figures depicted on the screen were. Above all, it is the experience of the earthquake's technological simulation that makes the most lasting impact.

The final exhibit, however, is where historical simulation, technologi-cal effects, and dramaturgy most fluidly merge. In it, a young twenty-first-century groom—one-half of the couple viewers meet on entering the museum—is seen rushing to his wedding by using the closest available transportation—a large, personal hang-glider. Launching himself from his bedroom window after realizing he's running late to the ceremony, he sweeps over long stretches of Lima's recognizable ocean coastline. He floats over the border where the city's urban edge meets a roaring ocean in a pitched techno-natural encounter, and touches ground in time to meet his bride at an outdoor *boda massiva*, or mass wedding, where dozens of other couples await. An array of evening festivities and fireworks flare across the screen as museum visitors witness the celebration. Although any number of endings for the film could have represented the complexity of Lima's his-tory, in all its fragmented and contradictory flows, it is this final setting that closes the narrative. In the background, spinning creole and pollera skirts of marinera and indigenous dancers mix, spiraling as the couples embrace and the bright, multicolored text, "Lima, ciudad para todos"—"Lima, city for all"—slowly comes into focus.

Despite being called a virtual museum, there is little in the way of inter-activity; visitors are invited less to actively participate in the telling of Lima's history than they are asked to pay silent witness to what it presents.

More surprising is what the museum does not show. There is no mention of some of the major twentieth-century changes that are part of the popular consciousness of contemporary Lima. No mention is made of the migrations from Peru's rural provinces that nearly doubled the capital's population between 1980 and 2010 (from 4.8 million to more than 9 million), and more than quadrupled the city's size since 1960 (when the population was estimated at 1.9 million). Such migrations are credited with turning the city from a once heavily white, European-identified capital of, almost exclusively, Spanish speakers to a city of mixed European, Indian, black, and Asian heritages where a number of native languages are audible.

Surprising, too, is the lack of any mention of Sendero Luminoso or the effects of the conflict on either the consciousness or physical spaces of Lima: the bombings in Miraflores, the sustained police raids on universities and working class neighborhoods, the emergency curfews and Fujimori's suspension of constitutional rights, or even the eventual capture of the Sendero leader Abimael Guzman in Lima. The oversight is all the more curious given how much the specter of Sendero and rurally sourced terrorism still shapes contemporary public discourse in Peru. Such concerns were expressed most recently, at the end of 2011, in the intense scrutiny and criticism of the Sendero-sympathizing collective, Movadef (Movement for Amnesty and Fundamental Rights) and its attempt to officially register as a political party. In the midst of Lima's new encounters with global markets and economic growth, the Movadef debate reminded the public of the risk that disconnection from global exchange could bring and the persistent threat of rural provincialism. Moreover, it renewed public indignation over citizens' lack of historical consciousness, the clear and continuing threat of terrorism and counterstate organizing in the provinces, and the need to advance global connection among rural communities where national memory and unity were, presumably, most at risk.

The promise of shared twenty-first-century memory, unity, and connection are, then, no small stakes. Little wonder that digital technologies, given their ability to enable past forms of global and local connection to be reimagined and recalled, gain such heightened value or inspire such diverse projects in contemporary Peru. Contrasted with the threat of local stagnation and supposed "resistance to progress" is the promise of a virtuality that can mobilize visions of the future across an array of sites. They represent diverse experiments with the digital that incorporate local and global dynamics and that pose more than the advancement of ever-newer, sleeker information technologies as their primary stake.

Signs of global connection, as anthropologist Anna Tsing (2004) has written to describe the contemporary, are indeed "everywhere." Far from exclusively pocketing in elite, hypernetworked districts of the world's cosmopolitan capitals, global connections, IT-enabled interfaces, and their associated seductions can be found as much among "remote" rural spaces as in glistening innovation districts. Still, it's precisely in the midst of global connectivity's growing ubiquity that demands for local connection and secure modes of engaging with distinct local cultures across stark divides grow more pronounced. Postcolonial theorist Gayatri Spivak's sage reminder that the universal is something we cannot help but not want, even as it excludes us, remains all too prescient (Najambadi 1991).

The mixed ambitions and anxieties inevitably drawn into what it means to pursue global connection, however, reveal themselves to be anything but uniform. How this has come to be the case in new millennial Peru, and how technology-centered social reforms have been distinctly and multiply imagined by diverse members of Peru's information classes is the subject of what follows.

Acknowledgments

This work would not have been possible without its own network of actors. Their work imparted innumerable lessons, and their words and activities never ceased to serve as unfailing guides. If this study has been able to illuminate a thing or two about the power of networks at the edges, it is a credit to the individual strengths and their countless quiet wisdoms from them it channels.

I am indebted to the many artisans of Chulucanas —in particular, Gerasimo Sosa, Santodio Paz, Cesar Juarez, Flavio Sosa, Narcisa Cruz Sosa, Emilio Antón Flores, Segundo Moncada Guerrero, Alex Calle, Juana Sosa, Jose Sosa, Luis Salas, Segundo Carmen, and Yuri Padilla. They opened the doors of their homes and workshops to me, and over bowls of caldo or cupfuls of clarito, they shared their stories and their art. It was my tremendous honor and pleasure to be in their company and to have been able to witness their work. I am tremendously indebted to the many artisans and organizers who have worked with the National Network for the Artisan's Law (RENAPLA)— and to Victor Zevallos, Walter Leon, and German Guillen, in particular. And my heartfelt thanks goes out to the residents and people of Chulucanas and the nearby villages of La Encantada and Ñácara who welcomed my stay, shared with me their trade and their homes, and were never anything but patient with the persistent questions of a curious young researcher.

My gratitude goes out the government officials and employees of Ministry of Tourism and Trade (MINCETUR) and the National Institute for the Defense of Competition and the Protection of Intellectual Property (INDECOPI) who allowed me to document their projects and travel in their company as they shuttled between Peru's capital and rural provinces hundreds of miles away. I am particularly grateful to Madeleine Burns, Luis Calderón, and their staff at MINCETUR and to Luigi Castillo, July Castillo, Mario Rubio, and their staff at the MINCETUR-managed nonprofit CITE Ceramicas based in Chulucanas. That they granted an independent researcher access to their offices and personnel in the process is a credit to them. My gratitude is likewise extended to Miguel Ángel Sánchez del Solar and the past and present staff of INDECOPI, including Juan Rodriguez, Ray Meloni, Martin Moscoso, Jose Tavera, Teresa Mera, and Nancy Matos. Thanks also need to be extended to Luis Cueva at the National University of Piura, and Jose Lecanos and Amancio Yamunaque at the Instituto Nacional de Cultura. And a very special thanks to Santiago Roca, whose work and leadership as the director of INDECOPI was an inspiration to many.

The many independent technical consultants who collaborated with INDECOPI and MINCETUR and lent their technical expertise toward these projects should also be acknowledged: Claudia Fernandini, Carmen Arana, Sonia Cespedes, Manuel Eguiguren, Augusto Mello, and José Dario Gutiérrez are foremost among them. I am especially grateful to the enormous intellectual generosity of Josefa Nolte, whose conversations about, and professional experience with, artisanal crafts in Latin America lent countless insights that I will continually revisit.

This work would not have been possible either without the invaluable work and contributions of the many activists, advocates, and programmers working through Peru's free, libre, and open source software community. This includes Escuelab's team of organizers in Lima—Kiko Mayoraga, Mariano Crowe, Juan Camilo Lema, and Raul Hugo—and its dedicated network of project organizers distributed throughout South America—Neyder Achahuanco, Irma Alvarez Ccoscco, Amos Batto, Vanessa Bosquez, Pia Capisano, Vladimir Castro, Aymar Ccopacatty, Koke Contreras, David Cruz, Javier del Carpio Lopez, Alfredo Gutierrez, Juan Carlos Leon, Daniel Miracle, Alex Munoz, Eleazar Pacho, Jonathan Rupire, Ivan Terceros, Michel Trujillo, Reyzon Trujillo, Sdenka Salas, Sebastian Silva, Laura Vargas, Matias Vega, and Walter Zamalloa. Their creativity and commitment never failed to surprise. And to Carolina Goyzueta, Andrea Naranjo, and Manuel Vargas for the generosity of their ideas—brilliant, wise, and always guiding. I am thankful, too, to Katitza Rodriguez, Pedro Mendizabal, and Katherine Cieza of Computer Professionals for Social Responsibility-Peru (CPSR-Peru), who lent their expertise and many a lively debate, even into the late hours of the night, on information technology and development in Peru. The friendship and support of members of the Peruvian Association of Free Software (APESOL), LinuxChix Peru, the Peru Linux Users Group (PLUG), Argentina's Free Path Foundation, and the global OLPC community was also invaluable: foremost among them Rudy Godoy, Fernando Gutierrez, Carlos Horna, Jesus Marquina, Antonio Ognio, Christian Peralta, Ernesto Quiñones, Rafael Salazar, Silvia Sugasti, Cesar Villegas, Carlos Wertheman,

Katia Canepa, Christoph Derndorfer, Bernie Innocenti, Federico Heinz, and
Enrique Chaparro.

I am enormously indebted to my committee members, Joseph Dumit,
Sherry Turkle, and Susan Silbey. Their intellectual generosity, sage guidance,
and unflagging support were extraordinary gifts that I will carry with me,
and I hope to pass on with the same tireless, ready flair that they did. It was
also my great fortune to be enriched by the talents, insights, and humor of a
number of colleagues and friends of MIT's Program in Science, Technology,
& Society and Comparative Media Studies Program, including Stefan Helm-
reich, Michael Fischer, Hugh Gusterson, Henry Jenkins, William Urrichio,
David Mindell, Pablo Boczkowski, Natasha Myers, Candis Callison, Richa
Kumar, Jamie Pietruska, Will Taggart, and Isa Dussuage. Lars Risan, Yuwei
Li, Sampa Hyssalo, Stephanie Freeman, Eve Darrian-Smith, Kitty Calavita,
Michael Chibnik, Manuel Guererro, and Jesus Elizondo provided insight-
ful readings, careful comments, and critical feedback on versions of these
chapters that unquestionably strengthened their content. Rachel Prentice,
Morgan Ames, Eben Kirksey, Tania Perez Bustos, and Manuel Franco Avel-
laneda are owed a particularly special thanks for the enormous generosity
of their ideas that enriched this work.

I am grateful for the support of the National Science Foundation's Gradu-
ate Research Fellowship, which provided funding for my fieldwork in Peru.
My warmest thanks goes out as well to Columbia University's Center for the
Study of Law and Culture for providing additional financial support for this
work and allowing me to benefit from the sage counsel and encouragement
of colleagues there: foremost among them, Elizabeth Povinelli, Katherine
Franke, Kendall Thomas, Tucker Culbertson, Bela Walker, Noa Ben-Asher,
Wade Wright, and Manissa Maharawal. A special thanks also needs to be
extended to the many friends and colleagues at the University of Illinois
whose thoughtful conversation and feedback pushed the development of
several chapters: Angharad Valdivia, C. L. Cole, Cameron McCarthy, James
Hay, Lisa Nakamura, Rayvon Fouché, Jodi Byrd, Christian Sandvig, Helaine
Silverman, Kevin Hamilton, Angelina Cotler, Markus Schultz, Safiya Noble,
Amanda Ciafone, Julie Turnock, Prita Meier, Harriett Green, Michael Sime-
one, Diana Mincyte, and Peter Asaro.

The wit, love, and support of a number of friends across space and time
have been an enduring source of inspiration to me: to Elaine Rivas, Anna
Tverskoy, Brandon Roth, thank you for these faithful gifts. And finally, my
mother, Catalina, and Roger and Rosalyn: for your continuing love and
humor, for the many lessons you've imparted, accidental and otherwise,
and for the many lessons still to come, I am inexpressibly indebted.

The 2010 ExhibePeru catwalk, staging new designs from indigenous artisans and Peru's ten rural centers for technological innovation (CITEs) in Lima's Barranco district.

Introduction: Digital Reform—Information-Age Peru

ExhibePeru's staging was otherworldly, even before the show began. Complete with an elevated outdoor catwalk, wrapped in red vinyl, and crowned by the cool, blue glow of a twelve-foot-high screen—it all made for a striking, futuristic theater, one that was all the more pronounced set against the late colonial architecture of Lima's Barranco district. With the hypnotically low reverberations of an electro track playing in the background, it was enough to make any spectator forget that it was a government agency—Peru's Ministry of Tourism and Trade—and not a sci-fi film that had been behind the event's organization.

For the past decade, the ministry has worked via its network of rural centers for technological innovation (CITEs) to demonstrate that Peru's traditional craftwork could be refit for the twenty-first century's global economy. Since their founding in 2002, the ten CITEs have focused on optimizing the "export potential" of artisans living in rural zones. This had been done primarily through encouraging new technologies that would transform traditional workshops into modern outlets for amplified craft production suitable for international tastes and demands.

Recently their work created a buzz at various state agencies when a project to develop IP titles for locally made ceramics in the northern town of Chulucanas began in the local CITE. If successful, the initiative promised to build recognition for craftwork as not only a physical expression of tradition but also a unique form of information work with its own innovation potential—and this, not just for Chulucanas, but for the nation's various rural and artisanal communities where the same model might later be applied. Indeed, since the mid-2000s, the Peruvian state had begun to promote the use of IP titles among rural producers, focusing efforts on the deployment of a form of IP known as a denomination of origin (DO) that would certify the production of regional goods, and developing nearly a dozen new titles for rural and artisanal products in the previous five years alone. Such productive activity allowed Peru to be chosen as the site of the World Intellectual Property Organization's (WIPO) 2011 World Symposium on Geographic Indicators—the first ever held in Latin America—which brought representatives from more than sixty-five nations into Lima to discuss recent developments in the use of denominations and appellations of origin as classes of IP.

As the lone product for which a DO had been developed that was craft-based rather than agriculturally derived, Ceramicas Chulucanas held a distinctive place in Peru's new IP pursuits. The state enthusiasm for that project and for the innovation-centered projects of the CITE network more generally were on display that particular evening. During the ExhibePeru event, fashion models slinked down the catwalk, exhibiting the updated, export-oriented wares of the rurally based CITEs—from textiles and metal weavings to, of course, ceramics—each refashioned to fit the cosmopolitan tastes of their urban-residing audience. The main stars of the show, however, were not so much models or even the government officials seated in the front rows but the rural artisans behind the crafts lavishly displayed. Working in sites hundreds of miles outside of Lima, it was these artisans whose active work and collaboration with local CITEs testified to the promise of entering global markets.

Their larger-than-life images were projected onto screens to document their work with raw materials—earth, wood, and stone—slipping between their hands. Some appeared on the catwalk at the end of a collection's display—the artisans' flared pollera skirts and bright, handwoven shoulder cloths, typical of women's dress in the Andes, flanked by the tailored suits and gowns of their escorts.

Following each collection, the CITE partners—artisans, designers, models, and all—disappeared from the stage. Then, as if to remind the audience that the whole event was one large programmed interface of the state's design, the screen behind them rebooted to the image of a hypersized computer desktop—one controlled by an invisible user's hand whose browsing launched the viewer's screen into a modularized digital overview of Peru's national territory. Once given the proper coordinates, it zoomed into the distinct CITE locale, breaching distances of air and land in a matter of a digital instant.

Without a doubt, multiple worlds, temporalities, and mediated fantasies all met on ExhibePeru's stage and seemed to melt into a curious and intoxicating evening brew. If there was any sense that IT and development, export and craft, or IP and native culture could be uncomfortable pairings, it wasn't betrayed there. The event channeled instead a conviction that innovation and tradition, information and material cultures, and rural and urban designs could be strategically coupled in mutually beneficial relations. The event followed announcements, too, that the CITE network would soon be expanded across four additional provinces (to a total of fourteen sites) and that projected exports were expected to help put Peru's GDP growth above 9 percent the next year (in defiance of global recession trends).

Still, miles outside the capital city, another far less pacific version of contemporary change was unfolding. Despite the nation's celebrated record of economic growth, disquiet and discontent persisted in Peru's rural provinces. At the same time as state agencies in Lima were celebrating the creative and economic potential of traditional populations, tensions involving indigenous communities in the north of the country—and new legislation to accommodate transnational gas extractions in the rain forests of the Amazon—were escalating. National newspapers filled with headlines about conflicts that led to the deaths of more than fifty protestors, indigenous community members, and security officers, numerous disappearances, and allegations that victims' bodies—as evidence of the violence—were being hidden (*El Comercio* 2009). The number of such confrontations had been increasing over the past several years as communities and labor sectors—from teachers to miners to farmers—in separate events across the nation were taking to the streets to decry the mismanagement of regional territories, and especially, the increasing presence and power of private interests in matters of local governance. By official estimates, the state was managing no fewer than 110 conflicts nationally at any one time since 2006 (Defensoria del Pueblo 2006, 2007, 2008; Peru Support Group 2007). That number would nearly double by the end of 2008, to some 216 cases, and would jump again to a record 293 cases by mid-2009, making images of marching demonstrators, blocked roads, and state-citizen confrontations regular additions to the nightly news. Especially in Peru's rural zones—where nature, land, and cultural difference remain objects of strident historical conflict and where state and corporate activities left a centuries-long trail of broken trust—tensions could be especially high. Without a doubt, political conflicts were mounting across the country, with populations' grievances against the state rarely confined to a single, easy category.

Back in Lima, however, the nighttime staging made for an undeniably entrancing setting. In the midst of other events, the ministry catwalk—bathed in the glow of fluorescent lights with Lima cosmopolitans and rural artisans walking hand in hand to the beat of a digital drum—provided one striking break from the rest.

Universal Network

However locally resonant the prospects of networked digitality framed at the CITE event—with its stress on harmoniously melded provincial and cosmopolitan interests—Peru has not been unique in its IT-based experiments. Indeed, at the beginning of the twenty-first century, the promise of digital connection is a projection that appears across an expansive range of transnational contexts that draw in ever-more distinct arrays of interests and investments. Once the near exclusive domain of a limited body of experts and enthusiasts—composed largely of Western computer scientists and engineers—computing and IT in just a few decades have become tools accessible to broad populations that span rural artisans and high-tech programmers alike. If in the mid-twentieth century few would have anticipated that many—beyond a small segment of technology specialists—could imagine the wide application of digital technologies, a decade into the new millennium few could reference popular conceptions of the contemporary or generalized practices of worldly connection at all without them.

Indeed, digital technologies today have come to be imagined as agents of reform whose impacts readily take effect as much at scales of the global as they do at the level of the individual. In accounts by technology experts and nonspecialists alike, they've come to be framed as everything from agents of global capitalism and accelerants of market integration (Friedman 2005; Ross 2004, 2010), to democratizing agents of knowledge exchange and information distribution once the exclusive domain of scientific or academic authorities (Lessig 2000; Willinsky 2009), to expedients of international peace, human rights, and democratic values that were once the lone purview of states and global governing institutions (Clinton 2010; Halliday 2011). Such potent characterizations tend to polarize orientations around digital technologies as obviously legible forces of change that push toward an already predictable future. Few, however, capture the range and complexity of engagements around digital networks that are indeed unfolding in a range of local sites—where networked undertakings suggest less a certain advancement towards a given pre-determined future, than an experiment in cultivating networked relations.

Google CEO Eric Schmidt [now Executive Chairman], for instance, has been apt to remind audiences that the digital age's current "information explosion" has already flattened and accelerated information flows so that all of humankind can now access as much data in forty-eight hours—five billion gigabytes worth—as had been created "between the birth of the world and 2003." Meanwhile, MIT Media Lab founder Nicholas Negroponte reminds the public that the coming "education revolution" and reform of failing education systems will come with digital technologies such as his organization's OLPC laptops. And global political leaders, including former US Secretary of State Hillary Clinton, stress that digital technologies will be forces of global political progress, asserting that "An open Internet fosters long-term peace, progress and prosperity." What the limitations of such predictions are remain unspoken; just as the notion of digital technologies as certain vehicles for universal transformation is framed yet again as a simple, taken for granted given.

Part of the reason such uni-dimensional future castings can remain in place, indeed, has to do with the relative scarcity of accounts we have of digitality—and so too, the relative limitations on who and what sites have been authorized to speak for digital presents and futures alike. Research on contemporary information cultures remains largely focused on a limited set of contexts—concentrated in Western centers of IT development and innovation—despite digital technology's oft-repeated capacity to usher in global reform. And while a redefinition of center-periphery relations and flattening of information flows have been purported as key effects of the global spread of IT and the Internet, relatively little focus has been placed on such spread across local sites. Research on digital networks has thus helped to develop keen understandings of information cultures and innovation environments in the West—looking in particular at the productive practices of locales such as Silicon Valley (Cringley 1996; Kidder 2000; Lewis 2001; Markoff 2006; Saxenian 1996; Turner 2006), technology-focused research centers and university labs (particularly those involved in developing the ARPANET) (Abbate 2000; Hafner and Lyon 1996), hacker collectives (Coleman 2011, 2013; Kelty 2005b, 2008), and IT enterprises that represent new forces of digital capitalism (Auletta 2010; Jarvis 2009; Vaidhyanathan 2011). But the diverse practices of adoption, and cultural aspirations and anxieties that emerge around digital culture's global geography have been far less documented. (Bowker 2005; Dourish and Bell 2011; Philip, Irani, and Dourish 2010; Takhteyev 2012) Even while it might now be recognized that IT engagements are no longer limited to Western IT experts and enthusiasts, it remains a relatively narrow set of voices that speak for an IT-enmeshed contemporary.

Of course, digital culture is not merely spoken for, or translated to broader audiences, by academic researchers alone. Key classes of highly visible digital enthusiasts—that include technology journalists, entrepreneurs, engineers, and designers, largely working in the West—have attained a spokesperson-like status to forecast the world's common digital future.

With messages of the universal power and promise of IT not merely conveyed in tech-centered magazines, blogs, and conferences (whether Wired, Techcrunch or TED), but increasingly the stuff of best-selling books and daily news cycles that target general audiences, too (Anderson 2006, 2010; Jarvis 2009; Johnson 2010; Kelly 2010; Rheingold 2003; Shirky 2009), an active digerati today have notably broadened the reach of their technological forecasts, and tailored predictions of a coming digital cosmopolitanism for new popular audiences.

When *Wired* magazine editors in the United States and Europe—Chris Anderson and Riccardo Luna, for instance—organized an online campaign to nominate the Internet for the Nobel Peace Prize in 2010 (Than 2010)—a move later sponsored too by MIT Media Lab founder and head of the global OLPC project, Nicholas Negroponte—they appealed to general publics online for support. Explaining the campaign on the English-language site Internetforpeace.com via a short video, the campaign organizers stated,

We have finally realized that the Internet is much more than a network of computers. It is an endless web of people. Men and women from every corner of the globe are connecting to one another, thanks to the biggest social interface ever known to humanity. Digital culture has laid the foundations for a new kind of society. And this society is *advancing dialogue, debate, and consensus through communication.* Because democracy has always flourished where there is openness, acceptance, discussion and participation. And contact with others has always been the most effective antidote against hatred and conflict. That's why the Internet is a tool for peace. That's why anyone who uses it can sow the seeds of non-violence. And that's why the next Nobel Peace Prize should go to the Net. A Nobel for each and every one of us. [emphasis added]

Although not emphasized in the missive, the original nomination of the Internet followed a series of protests in Iran to contest the controversial 2009 presidential victory of Mahmoud Ahmadinejad. Reporting in Western media outlets—from the *New York Times* to the BBC and *Wired* magazine—stressed the use of digital networks by protestors, whose online broadcasts helped draw hundreds of thousands of fellow citizens out to rally and supplemented global news coverage beyond. Eyewitness accounts streaming across YouTube and Twitter evidenced, as one BBC headline stated, how the "Internet [brought] events in Iran to life" (BBC 2009b). In the weeks following the protests, Mark Pfeifle, the US deputy national security adviser at the National Security Council in 2009, issued his own public plea that Twitter receive a Nobel nomination (Pfeifle 2009). The sentiment was echoed by *Wired* magazine Italy editor, Ricardo Luna, who explained the 2010 Nobel campaign by triumphantly declaring the Internet as "the first weapon of mass construction which we can deploy to propagate peace and democracy" (Wallace 2009).

Such fervent projections around the universal promise of digital connection, of course, draw upon the notion that participation in digital networks can by itself activate the Internet's core technological principle as a "tool for peace." Such a notion turns on the idea of the Net as a uniquely and universally inclusive and equalizing space—one that operates free from the physical biases of the offline world. Regardless of the particularities of individual embodiment or local condition, in other words, once online, *all* users could be granted the same agencies on a single network, all differences could dissolve, and everyone could be treated alike as "just users."

Much, however, is left unsaid in the missive. Directed toward "anyone who uses" the Internet, for instance, the "Internet for Peace" manifesto omits any reference to local specificity, appealing instead to a generic and explicitly placeless cyber-populace. Despite its opening that the Internet is "more than a network of computers" and, indeed, is "an endless web of people," it issues a strikingly anonymizing, context-erasing account. It makes no reference to the critical events that led up to the Nobel nomination, just as there is none made to the protestors who faced off with national security forces that past year, or to the history of local efforts and struggles that preceded them, or even to the individual bloggers whose reporting, in spite of local censorship and on-the-ground risk, had brought them to the attention of Western media. The campaign instead collapsed such distinctions, pronouncing the Internet as the same free and common digital meeting ground that audiences already recognized, without need for further geopolitical familiarization. Playing on the slippage between ordinary, individual action online and evident history-making transformations, it credited the shared spirit of all net users, from whatever corner of the world, for the events of global revolution. Whatever the manner or content of one's online activity, "each and every one of us" could be deemed collaborators in the extension of a unified global digital peace.

Still, that such declarations of collective peace originating in Western centers could adequately speak for the variety of locally anchored concerns and ambitions that inform digital practice, has begun to look ever-more untenable in recent years. In the wake of the global spread of networked protests—from the Arab Spring to the European Indignados and international Occupy movements—the need for analyses developed outside of given and established technological centers has become increasingly evident. Prior to the popular uprisings in the Middle East that compelled multiple governments to issue Internet and cell phone blackouts in the early months of 2011, few—particularly within innovation's established centers—would have considered places like Tunisia, Egypt, or Syria to be critical to understanding the Internet. Afterward, few could deny they were. If studies of digital culture once took for granted that a focus on centers of technological design and development would be adequate to sufficiently predict global transformations, the surprise of events occurring at the so-called periphery of innovation proved otherwise.

This project shifts the Western focus and techno-universalist assertions in studies of the Internet as a global network to explore new ICT-based pursuits at the so-called peripheries of technological innovation—in Peru, a South American nation far less attended to for contemporary innovation than for its ancient Incan ruins and high Andean peaks. Since the new millennium, however, a growing number of events have distinguished Peru's experiments in ICT deployments. This includes recognition by global governing bodies—from the UN and WIPO—for its aggressive pursuit of IP titles that aim to turn rural populations into new "creative classes." These novel efforts provide the focus of the first section of this book considering state-based visions of ICT in the making of the nation's digital future. Other events that have distinguished Peru's experiments in ICT deployments include the introduction of one of the first legal proposals to obligate state use of FLOSS, and establishing the largest global partnership with the MIT-derived OLPC program, which has distributed almost one million laptops to largely rural schools. Both cases are treated in the second part of this book on citizen-based interventions around ICTs that are likewise invested in the making of digital futures, but that push for *other* forms of digital connection and relationality.

Such cases demonstrate that even when Peru engages issues that prominently feature in Western discourse around digital culture—including IP, hacker practices, and the selective logics of the information economy—in introducing elements underaccounted for from the vantage of innovation centers, it also exceeds such dominant frameworks and reveals them as simply one possibility among multiple, prospective others. These "achievements," however, are not the primary reasons to draw attention to ICT-based developments at the Peruvian "periphery." If such cases depart from the given models of IT-based innovation and design, their example should not be heeded as a means of simply gaining a better diagnostic on how emerging sites of dominance arise or to understand why former centers might get displaced. Although generated from populations who have been among the historically marginalized and underrepresented, after all, the new directions taken by such actors aren't necessarily acts of virtuous resistance to existing models of domination (Robbins, 1998). Much like pathways for digital connection architected via technological centers, they bear the possibility of establishing new means to exercise control over collectives as much as they generate possibilities for countering them. Both operations demand new attention from scholars.

Pursuing Connection

Launched to explicitly promote cross-sector, public-private collaborations at the global scale, state-run "information society" initiatives reflected one means of articulating new, nationally-anchored zones of ICT-enabled exchange. In organizing the first 2003 World Summit on the Information Society (WSIS) meeting in Geneva and the 2005 meeting in Tunis, the UN stressed the importance of building a "multi-stakeholder process" and expanded its pool of participants to formally include IT-focused businesses. Only the second time in its history that the UN accredited businesses for participation in a summit, it enabled the number of participating businesses to grow from 98 to 226 between WSIS's 2003 and 2005 meetings (surpassing the 174 states represented in 2005).[1] Calling such multisector presence "essential" to building an "inclusive Information Society" in the 2005 Tunis agenda, it underscored the importance of "strengthened and continuing cooperation"[2] in "public private partnerships" at the national, regional, and international levels.

Peru's response was immediate, making plans within months of the first WSIS in Geneva to establish a national, multisectorial Commission for the Development of Information Society (CODESI). Overseen by the newly founded National Office for Electronic Government and Informatics (ONGEI), it drew together more than two hundred experts from nearly ninety public, private, academic, and civil society institutions (Ferrer 2009). Within two years, their efforts produced Peru's plan for the development of the information society, a 140-page document known as the *Peruvian Digital Agenda.* Pledging to propel the nation into the information age, it opened with the thrill and promise of this opportunity, stating, "The 21st Century brings Peru and the world face to face with the Information Era, where . . . potentials for establishing a society based on access to information and knowledge . . . are already coming to revolutionize the state, law, the economy, and society" (CODESI 2005, 3). It specified, however, that such change would not be driven by government action alone but would require "the valuable contributions of civil society and the private sector." And it indicated "the Information Society doesn't emerge through the mere will of participants—but through the . . . execution of specific policies by diverse public and private actors." (CODESI 2005, 3)

The language of connection making had other implications for national populations. The new emphasis in international governance on the "knowledge economy" and "information society"–based projects fostered initiatives to promote not only information-based resources but also new "competitive" skills to put them to use. These included the use of networked IT (computers, mobile phones, and web-based resources) and IP rights for new, information-based "goods" exchangeable through such networks (Dutfield and Suthersanen 2008; Finger and Schuler 2004; Krikorian and Kapczynski 2010; UNCTAD-ICTSD 2003; Unwin 2009; Wong and Dutfield 2010; World Bank, 2009). Underscoring how such technologies promote the cultivation of competitive skills and competencies by individual actors, Peru's 2005 *Digital Agenda* cited the work of prominent network society theorist Manuel Castells to emphasize ICTs as forces of social and individual transformation. As it summarized his arguments from the first volume of his *Information Age* trilogy (1996), "Productivity and competitiveness in information production is based on the generation of knowledge and processing of information. Knowledge generation and technological capacity [today] are key instruments in the competitiveness of businesses, organizations, and countries." The challenge Peru's 2005 Digital Agenda made clear then, would be to apply such principles not only within urban populations and zones but among rural ones dispersed across the nation too.

Six years later, CODESI again emphasized the growth of the nation's ICT infrastructure in a new report. It cited the rise in the number of cell phone lines from seven million in 2006 to twenty-six million in 2010, with more than 80 percent of the country having complete or partial signal coverage; a growth of fixed telephone lines of 25 percent; and new state investments of $100 million USD in rural Internet and bandwidth services in 2010 and 2011. And although some 6.9 million people, or 27.4 percent of the population, had some form of access to the Internet in 2008 (largely through local Internet cafés), the number of home subscribers increased from 2 percent to 12.9 percent between 2008 and 2010 (CODESI 2011; Ferrer 2009). In the space of just a few years, rapid changes in the distribution of ICTs turned global connectivity into new "everyday" experiences for rural and urban citizens alike. The 2011 CODESI report explained the transformations: "That ICTs are [now] intrinsically linked with the routine and daily actions of a significant percentage of citizens around the world puts the greatest means of communication and interaction and development within our reach today." It went on to say that contemporary conditions of "economic

globalization . . . [and] a growing interdependence among countries" neces-
sitated new policies to build a "person-centered" and "development-ori-
ented" information society, "where *everyone* can create, access, utilize and
share information knowledge, and where communities and peoples can use
fully their potential in promoting sustainable development and improving their quality of life." Although it specified that such principles would
echo UN charters—such as the Universal Declaration of Human Rights and
Millennium Development Goals—it also stressed the importance of "innovation," with ICTs as a means of heightening national economic competitiveness. The report asserted unequivocally that the "popularization of ICTs
offers opportunities for new waves of innovation that as a country, we must
seize" (CODESI 2011, 37).

Such aggressive goals have characterized several of Peru's ICT-centered
ventures targeting rural populations in particular, including new work to
develop IP titles for "traditional" cultural and agricultural producers. In the
past several years, Peru had distinguished itself internationally by developing nearly a dozen new DO titles for rural and artisanal producer communities. Until 2005, Peru had just two DOs—one for the brandy Pisco and
another for Cuzco's maiz gigante—but since then, the number of active
DOs has quadrupled. Following Chulucanas's DO, new titles were issued
for coffee from Villa Rica and Huadquiña, pallares beans from Ica, loche
pumpkin from Lambayeque, maca tubers from Junin, and cherimoya from
Cumbe (with five more in stages of processing as of this writing). Legal
scholars meanwhile have questioned the efficacy of IPs as instruments of
contemporary rural development and democratic change (Boyle 2010; Correa 2000; Dutfield and Suthersanen 2008; Krikorian and Kapczynski 2010;
Matthews 2002; Sell 2003; Wong and Dutfield 2010).

The DO initiative is not alone in Peru's expanding state-sponsored
experimentations with IT. Peru was one of the first nations to partner with
MIT's high-profile OLPC project in 2007. Although the national Ministry of
Education had deployed digital education programs since the early 2000s,
its partnership with the OLPC Foundation—one that remains OLPC's
largest, with nearly one million laptops distributed and more than $280
million USD committed to date for the project in Peru—gave the effort
new global visibility. Even without large-scale studies to confirm the pedagogical impacts of integrating laptops in classrooms, and in rural zones
in particular (Warschauer and Ames 2010), the program rapidly expanded
from pilot to national policy in Peru—with OLPC's miniature green laptops
appearing in a matter of months in thousands of classrooms, newly labeled
as "innovation classrooms."

Such heightened investments in ICT-centered programs that stress the productive potentials of IT and information-based properties as national resources underscore the essential role of global networks for their deployments. But so, too, do they demonstrate the crucial element of local connectivity. They are based not only on the relationships that emerge from global collaborative partnership but also on those sustained by local ties. These initiatives require, in addition to national government agencies, transnational exporters, foreign IT companies, international NGOs, Western universities, programmers from around the world, participants from local communities, public schools, and traditional producers.

It is the newly intensified role for individual participants—and the redefinition of marginal actors as new agents of reform—that distinguishes ICTs from technological solutions extended under earlier development paradigms (Escobar 1994; Ferguson 1994; Mitchell 2002). ICT solutions become especially novel for contemporary development in precisely the way they invite equivalent adoption practices among diverse populations—from urban knowledge workers to the very rural communities that are guardians of traditional culture. (Coombe 1998a; Dehart 2010; Hayden 2003) They are useful that is, in the way that they can invite new productivities to be extracted from urban and "enlightened" citizens as well as rural and indigenous populations. In a single uniting sweep they draw on the creative energies of those taken to be exemplars of—and those once only seen as targets of—modern reform, promising to allow diverse actors new opportunities to channel their intellectual efforts. And as if to attest to the radically transformative potentials channeled through ICT networks, subjects who had once been held as the abject other to modern universalizing projects could now be given the chance to emerge as model global participants instead.

The chapters that follow examine the local particularities in the global connections that underpin the expansion of ICT's sociotechnical networks. In Peru, such networks have not only depended on the distribution of new information-based technologies and resources but have also involved commitments from contemporary government planners, transnational corporations, rural community members, traditional artisans, and information technology experts and engineers. If ICT networks have advanced universalizing ambitions, it is in part because they enable the building of strategic alliances between urban and rural subjects, the high-tech and the traditional, the wealthy and the economically marginalized between the social and technical and between the cultural and natural. They have encompassed not only actors who have traditionally held power but also those who have been denied it. And they have demonstrated this transformative force, not only by assembling new rural innovation spaces and remote zones of information production (what might be seen as parallels to urban forma-tions of global cities and enterprise zones) but also in the very remaking of citizens themselves—including the poor and economically marginal—who can now be repositioned as optimized and even hypercompetitive global performers.

ICT's universalizing expansions, and the notion that they function to extend new opportunities, then, should be understood in relationship to earlier universal projects, and in particular, to the limitations and liabilities of those earlier projects. Postcolonial scholars brought recognition to the global spread of multiple exclusionary logics of Eurocentric modernity—visible in not merely the global spread of colonial exploits but also in the spread of the notion of European domination as the natural expression of superiority over biologically inferior and culturally primitive others (Dus-sel 2000, 2002; Quijano 2000, 2007; Said 1979; Wallerstein 2006). Many science studies scholars have critiqued scientific universalisms for exclud-ing populations for a lack of rationalism, unenlightened and naïve belief systems, or fundamental inabilities to discern myth and fetish from fact and truth (Latour 2003; Stengers 2007; Viveiros de Castro 2004; Wallerstein 2006). Digital networks, however, offered an alternative. Even if they might not undo or extinguish the exclusionary effects of prior Western universals, they could produce new logics of inclusion offering the formerly excluded a means of participation and even competition within global networks.

Such distinct operations of inclusion are critical, then, precisely in their seductive force and contrast to the past exclusions of other Western universals. Similar to contemporary networked assemblages whose socio-technical inclusions accommodate collaboration among diversely situated actors (Benkler 2006; Callon 1986; Callon and Latour 1981; Castells 1996; Hayden 2003; Latour 1987, 1990, 2004a; Mitchell 2002, 2005; Ong and Collier 2004; Riles, 2001; Weber, 2004) and enable such varied interests to be drawn together for the work of reform, digital networks extend the promise of new connections between the center and periphery. Central to the conditions that enable (or might disrupt) ICT extensions from taking place as rapidly as they have are the particularized "situated" experiences of networked actors (Haraway 1991), who make possible the building of alliances in a growing diversity of sites—and among even the most unlikely of partners. What brings such partnerships together and seduces such dis-parate actors into alliance is not only the promise of increased access to technology, information, or other material resources but also the promise of new forms of local and global connection—including the opportunity for diversely situated subjects to realize themselves as future-oriented, cross-nationally networked individuals freed from the oversight or intervention of established institutional and political actors. Said otherwise, they may be inspired by what political theorists Nikolas Rose, Andrew Barry, and Thomas Osborne (1996) write of as a "'form of politics beyond the State, a politics of life, of ethics, which emphasizes the crucial political value of the mobilization and shaping of individual capacities" (1996, 1)—but now, with IT-based technologies supplying new means by which such capacities as creative and intellectual forces can take global flight.

The chapters that follow explore the multiply scaled relations that unfold for newly networked participants working around the prospect of global connection through investigating ICT-based initiatives in Peru. Chapters 1 through 3 consider the forms of global connection engineered through the Peruvian state's experiments with export-oriented intellectual property initiatives. Such programs have targeted rural and traditional producers as emergent classes of creative workers. Much as the creative classes and knowledge workers concentrated in hyperindustrialized global cities, urban innovation zones, and technology clusters are framed as sources of cultural and intellectual innovation in the industrialized West, here, too, traditional producers are hailed as sources of expressive work and cultural innovation that now lend fuel to the Peru's information-based economy. Focusing on the collaborative formations that have produced the IP title for traditional ceramics from the northern coastal town of Chulucanas, then, the chapters highlight interviews with the government planners, global exporters, and development experts who have hailed IP's capacity to amplify the global flow of "native products"—as well as with the traditional artisans whose production practices are indeed radically remade in the interest of "optimizing" capacities for global competition and information property management. Chapter 1 thus follows the local relations formed in Chulucanas, extending scholarship on network formations that have emphasized the means by which they work to enroll participants and recruit allies in newly inclusive acts (Benkler 2006; Castells 1996; Lessig 2002; Mellucci 1996; Ong and Collier 2004; Sassen 2002, 2006; Weber 2004). It demonstrates, however, how such "successful" network incorporations may simultaneously function as a disintegrative force, dissolving or weakening the bonds of preexisting social relations that sustained the work of cultural expression in local communities and exercising a selective logic of incorporation into global partnerships.

The following chapter highlights the means by which ICT-based development programs provide platforms that allow state actors to signal new valuations around local cultural diversity and performances of "authentic" cultural traditions. Chapter 2 thus explores the ways in which IP initiatives allow the Peruvian state to newly claim a role in protecting provincial traditions and native cultures, while simultaneously promoting rural producers' "export potential." Although the state publically celebrates a message of "cultural preservation" and protection of shared national heritage through its IP initiatives, in practice, they selectively reward local producers based on an ability to demonstrate "competitive" potential and to execute reforms on traditional production practices. Much as information economies have been celebrated for unleashing the creative potential of individual producers (Florida 2003; Howkins 2002), so, too, have Peru's IP-based initiatives granted new roles to once-excluded traditional producers, who can now be recast as heroes in the work of promoting national tradition. And in the process, they allow the state, too, to be recast as a new version of itself as a 21st century, multi-cultural steward.

How earlier Western universals and the modernizing promise of global markets still haunt rural IP initiatives is the topic of chapter 3. Here, international exporters' insistence on the power of "modern" export practices to transform rural producers into "gentlemanly" businessmen, who are at once fit for negotiations in a global economy, figure centrally. Much as historians of science have pointed to the "gentlemanly" and "modest" dispositions European patrons demanded from the scientists whose endeavors they'd fund (Biagioli 1993; Shapin 1995; Shapin and Schaffer 1985), so, too, do global exporters insist on "modest" dispositions from artisans seeking export partnerships. Although pro-export actors appeal to the rational "evidence" of the modern market's civilizing promise, local artisans spin another set of tales. Against IP and export promoters' salvation stories and master narratives of rural advancement and new global competitiveness, then, are accounts that describe global partnership and export-based networks as cultivating relations of promiscuity with diverse parties—including those with oppositional interests to cultural or community preservation.

The logics and seductions of international export, however, are not the only ones that determine relations of global connection or the form they take via new ICT networks. If the state's IP-based pursuits represent a means of integrating Peru's cultural producers into global circuits of economic exchange, other cultural developments demonstrate efforts to question, reprogram, and disrupt the frictionless integration of new populations into such global market-oriented channels and insist on alternative forms of locally grounded relations through global connection. Chapters 5 through 7 explore community-innovated alternatives to state-deployed information society programs by investigating Peru's FLOSS advocacy networks. Although the growth of FLOSS communities was hailed in US and European technology circles for innovating a model of distributed, volunteer-based software production, Peru's FLOSS network came to the attention of global publics for deploying a model of distributed, volunteer-based, legal production with an explicitly political charge. Their activities gained visibility in the mid-2000s when Latin American FLOSS activists helped Peru become one of the first nations to propose legislation for the statewide adoption of FLOSS-based technologies. To advocates in Latin America, such open technologies offered a powerful means not only to critique the hegemonic power of Western technology corporations in domestic markets but also provided a means for arguing for broader participation from diverse sectors of civil society. Seeking to cultivate an ethic of social inclusion and broadened participation, FLOSS advocates believed it important to operate as a network of cultural—and not only technological—innovation whose activity could destabilize established political practice and meaning and generate new imaginaries around what could be adopted as new political norms, practice, and policy.

Peruvian free software activists have also undertaken collaborations with rural teachers for experiments in digital education projects that develop independently from the state's own digital education initiatives. Working around local deployments of OLPC, the subject of chapter 7, such emerging partnerships build new cultural interfaces between rurally based teachers and engineers, indigenous leaders, Quechua and Aymara language activists, and global information activists. These convergent efforts demonstrate how digital politics begin to unfold in rural spaces, drawing in diversely situated global actors and interests that, nonetheless, cultivate local politics of place. Such collaborations create instances of network "interruption" that unsettle and counteract the interests of the Western-based IT entrepreneurs or Peruvian state authorities who dominate global channels of resource distribution around digital education projects like OLPC.

Together, such chapters account for a cultural condition of ICT futures that are still, literally in formation. Although these networks serve the interests of established centers of political and scientific power, so, too, do they channel the interests of regional information activists, local and rural knowledge workers, and citizens with varied technological interests and ambitions. Their encounters draw together globally circulating expertise and knowledge with locally cultivated parallels that bear the possibility of reorienting engagements with local spaces and allow a distinct ethics and situated practice of global digital networks to come forth. Here, it is not so much the measure of individual enterprise or networked competition that marks global agency. But rather commitments to cultivating encounters of genuinely engaged connection across difference, through means that seek anything but to convert and assimilate the other, are projected as holding the greatest promise (Stengers 2007).

Peace, Love, and IT Affairs

Of course the forms of engaged digital encounter-across-difference that Peruvian FLOSS and rural technology activists have aimed to articulate as paths towards a distinct kind of future is far from the only means that networked relationality has been imagined. Indeed, the Internet for Peace Manifesto promoted by key actors within the global digerati betrays another kind of projection for networked relationality—one where a borderless, peace-oriented globality could be achieved via the self-organizing technological operations of the Net, rather than the contextually-conscious, locally-deliberated commitments and negotiations Peruvian activists insist upon. The technical, self-automating architecture underlying the Internet matters here precisely because it seems to function under the direct control of individual users. It can thus appear to self-administer as an authentic reflection of the unmediated, collective will of users who can freely interact without the interventions of external collectives or institutions. Whatever problems of "representation" might emerge can be assumed to be either technical or simply an issue of a (resolvable) lack of information rather than an entrenched problem rooted in histories of social exclusion and political marginalization. Not unlike framings of twentieth-century development techniques that critics of neoliberalism cautioned against (Escobar 1994; Ferguson 1994), contemporary technologies can be summoned as apolitical solutions oriented toward universal progress for problems seen to have been initially generated and continually exacerbated by traditional politics.

Of course, the digerati's embrace of the Internet as a politically neutral expedient for global peace has not been without contention. The celebratory pronouncements of networking technologies as fundamentally promoting individual liberties and, hence, democratic and anti-authoritarian practices internationally have been critiqued for the "cyber-utopianist" zeal uncritically adopted by Western policy makers, engineers, and IT-focused enterprises alike (Carr 2010; Lanier 2010; Mackinnon 2012; Morozov 2010a; Turkle 2009, 2010). Stressing their "naïve belief in the emancipatory nature of online communication" the Belarus-born technology policy analyst Evgeny Morozov labeled such faith as a "net delusion" shared among global digital elites (Morozov 2010a) that neglected the pressing specificities of political and social context.

Although the digerati's projections of pacific, cosmopolitan unification are ambitious, they share earlier roots. Generally framed as a "program for the containment of global aggression and the universal respect for human dignity" (Nussbaum 2010, 28) cosmopolitanism's intellectual tradition is one philosophers trace back from Immanuel Kant to the broader ideals of the Enlightenment and the philosophy of the Greek stoics (Beck 2004, 2005, 2010; Habermas 2010). Martha Nussbaum (2010, 29) points in particular to the stoics' concept of becoming "citizens of the world" as an early projection of collective association that surpassed state-based citizenship. Kant's outline for extending cosmopolitan principles—optimistically asserted in his famed treatise on "perpetual peace"—meant to resolve the specter of war that he feared was mankind's greatest liability. Globalization theorists today, however, note that Kant pointed to the material conditions for fostering cosmopolitanism's "world political community" and "universal state of mankind" as already present in eighteenth-century Europe's global extensions—namely, in international circuits of commerce, fine arts, and early scientific culture (Cheah 1998; Cheah and Robbins 1998; Robbins 1998), that he read as incompatibile with states of war. They stress too how the legacies of such ideals manifest in the modern political architectures that encompass global governing bodies like the United Nations (UN), visions for a common political community embedded in liberal individualism and shared legal norms in the administration of international law, and the spread of global standards for human rights, environmental regulation, and IP (Cheah 1998; Cheah and Robbins 1998; Robbins 1998). They manifest, too, in more everyday spaces through the localized application of such norms and the local interventions of varied global justice movements and humanitarian efforts (Brown and Held 2010). And yet, popular faith in the ability of such institutions to produce a self-extending, borderless "perpetual peace" seems more than ever to be compromised, with a loss of faith in the operations of established state institutions and bureaucracies assuming growing populist guises.

In the United States, contemporary movements from the conservative Tea Party to progressively oriented Occupy movements have adopted networked formations in their critiques of state power. Science scholars have noted, too, that modern institutions of science, higher learning, and research likewise have had their own credibility crisis (Ferris 2010; Mooney 2005; Oreskes and Conway 2010; Pooley 2010) on issues from global warming, to environmental policy, and health monitoring. Such growing disillusionment in the West with established institutions of science, education, and liberalism that represent prior instantiations of universal ideals, however, seems to have left the Net's version of universalism virtually unscathed. If the rapid growth and popular uptake of networked digital technologies provides any indicator, digital universalism appears to have done what older universal projects have been unable to today—to convince ever broader audiences of its incontrovertible, self-evident, and universally shared necessity.

The civic interventions around and reactions to ICT futures emerging from the periphery, however, diverge from such an unconditional embrace of a self-evident digital necessity. To point to such cases offers up more a means of posing questions rather than providing a particular answer. Through them, a query opens up around the possibility of cultivating and multiplying other forms of "universal" ties and activating ICT-based connections around the global. It's in such operations that the prospect of what postcolonial and science studies scholars alike have framed, in posing critiques of modernity's universal forms, as "engaged universals" emerges instead (Chakrabarty 2000, 5; Tsing 2004, 8). Such engaged forms would acknowledge the promise and peril they carry as bodies of "knowledge that move," like all universals, across local space, time, and cultures (Tsing 2004, 7). Rather than take for granted the uniform replication of their features as they travel, however, they could instead insist on the weaving together of new "historical conjunctures" and "relational rapports" (Stengers 2011), interlinking sites of experiment as they travel. As such, they represent the promise of opening new channels of exchange between zones of encounter. So, too, however, might they foreclose the possibility of other imaginaries if not careful of the potential to act as assimilating forces.

This project thus begins by unpacking the diverse imaginaries and innovations that underpin digital culture and that surround networked connections—organized in and around new information and communication technologies —as they take on global dimensions from the margins. It is a project that explores a notion becoming increasingly evident a decade into the new millennium: that to follow the creative classes, knowledge workers, information activists, and new creators of IP as they carve out pathways and circuits at the edge is not to travel the same path as before, even when in conversation with the "universal" terms anchored by Western debates and interests. And it is also to discover actors contending with those who assert there is a single, universal path of networked culture and a "global information society." Indeed, the dispersal of ICTs has given actors at the periphery new capacities to pursue their own means to global connection and networked agency. Notably, their experiments don't simply reproduce models from Western centers but take on their own, unpredictable contours.

1 Neoliberal Networks at the Periphery

A street view of Chulucanas taken by the author during sunset in late 2006, the year Chulucanas was awarded the nation's first IP title for craftwork. The exterior of several artisans' workshops and homes can be seen in the shot.

1 Enterprise Village: Intellectual Property and Rural Optimization

Settled in the dry, dusty, coastal desert of northern Peru lies the remote, sun-washed town of Chulucanas. Throughout Peru, it is known principally for its arid climate, its green stretches of mango and lime cultivation, and for a sizeable and (now more than ever) growing community of traditional ceramics artisans. Adopting techniques derived from the pre-Columbian Vicús and Tallán cultures that populated the area thousands of years ago, Chulucanas ceramists see themselves as the inheritors and guardians of indigenous traditions in handmade ceramics—traditions that include stone polishing, paddle molding, and clay smoking with mango leaves. It is this rural province that has become the unlikely site of investment and experiment for the Peruvian government's contemporary information society initiatives, where the promotion of IP rights—and the cultivation of authors and inventors as IP rights holders—in the interest of economic development and regional modernization figure centrally.

In 2006, Peru's government launched a project to promote the use of an IP title—known as a denomination of origin (DO)—in the more than four hundred ceramics workshops in Chulucanas. The geographic indicator would serve as a kind of location-specific brand that identified Chulucanas as the exclusive site of origin for the ceramics and that attributed its particular characteristics to its geographic roots. Acquired in mid-2006, Chulucanas's IP title was hailed by the government as a means of securing multiple economic and cultural benefits for Chulucanas. First, it would serve a branding function for consumers internationally that would distinguish Chulucanas ceramics in the global market—and might thus promise similar benefits as Champagne's geographical indicator had allegedly yielded for champagne producers in France. Second, it would establish a set of "modern" market-oriented standards and regulations that ceramicists would have to adhere to in order to qualify their products. And last—and especially critical—it would recognize and protect the ancestral traditions of ceramics making inherited from the Vicús and Tallán peoples dating back to 500 B.C. As the Peruvian state's promotional guide for the DO, published by its National Institute for the Defense of Competition and Protection of Intellectual Property (INDECOPI), explained, such titlings should function to "*distinguish and protect* a product," basing such protective functions "upon the special characteristics essentially derived from the geographic environment in which [the product] was made—including natural, climatic, and human factors" (INDECOPI 2010, 4 [emphasis added]).

Chulucanas has not been the only zone of rural production to be distinguished with an IP title by the Peruvian state in recent years. Indeed, since the mid-2000s, Peru's government offices have begun to newly promote the use of IP titles among rural producers, focusing efforts on the deployment of DOs that could certify the production sites of regional goods and developing nearly a dozen such new titles for rurally and artisanal products. Such intensified activity was what allowed Peru to be selected as the site of the WIPO 2011 World Symposium on Geographic Indicators—the first ever held in Latin America—that brought representatives from more than sixty-five nations into Lima to discuss recent developments in the use of DOs as classes of IP. The event, without question, channeled a shared enthusiasm for the heightened potentials of IP applications among rural producers as potential classes of new knowledge workers. Given the Peruvian state's efforts succeeded, after all, its initiatives promised to build recognition for craftwork as not only a physical expression of tradition but also a unique form of information work with its own innovation potential—and this, not just for Chulucanas, but for the nation's various rural and artisanal communities where the same model might later be applied.

Responsible for promoting the denomination of origin as a strategy of regional development and modernization were two government offices: Peru's MINCETUR and the national office responsible for managing IP rights, INDECOPI. These two branches of government worked with Chulucanas artisans to ensure their participation, coordinate government representatives in Lima and provincial sites, and maintain ties with exporters working with Chulucanas ceramics. It might seem that the state's interest in Chulucanas as a contemporary site of cultural expression and national productivity would generate a new collective consciousness and communal identification among local artisans—particularly given that building collaborative partnerships between the diverse actors invested in a region has been a defining strategy of the state IT-based projects. Surprisingly, however, it seems to have in fact produced precisely the opposite effect—disintegrating shared public interest and dissolving spaces of collective identification. What I heard from artisans were not testimonies of shared interest and community collaboration but instead accusations of betrayal against other ceramicists, talk of design stealing, accounts of price warring and wage exploitation, and expressions of general distrust of the public institutions and civil associations that failed to put an end to such practices. There seemed to be a notable disintegration of shared public interest and collective identification. One could say that contemporary relations among artisans following the promotion of Chulucanas as a productive site of culture as intellectual property are characterized for their pointed suspicion, competition, envy, and even exploitation.

There is little sense of this conflict, however, in the flurried pace by which intellectual property is being newly—and by some registers, surprisingly—incorporated into development policies by nations worldwide. Since the World Trade Organization's adoption of the Trade-related Aspects of Intellectual Property (TRIPS) Agreement in the mid-1990s—which underscored IP's role in global economic productivity and stressed the need to standardize international protections—nations, rich and poor alike, have accelerated efforts to develop and register IP rights and to strengthen the legal infrastructures for national deployment and global coordination (Correa 2000; Drahos and Mayne 2002; Howkins 2002; Mascus 2000; Matthews 2002; Sell 2003). Such trends coincided with new emphasis in development circles on "knowledge economy"—and "information society"—based projects for new development initiatives. Promoted by organizations including the World Bank's Information and Communication Technology Sector Unit, UNESCO's Communication and Information Sector, and the World Summit on Information Society forums held in 2011, 2005, and 2003, the focus on "information society" reforms fostered the growth of initiatives to encourage populations to develop new "competitive" skills around information-based resources—including the use of IP rights (Dutfield and Suthersanen 2008; Finger and Schuler 2004; Krikorian and Kapczynski 2010; UNCTAD-ICTSD 2003; Unwin 2009; Wong and Dutfield 2010; World Bank, 2009). The World Bank and WIPO especially endorsed national pursuits and the dominant framing of IP as an economic utility, reasoning that expanding IP systems can "reward creativity, stimulate innovation and contribute to economic development while safeguarding the public interest" (WIPO 2011). Scholars meanwhile observed how the economically generative power of IP had powerful effects on government decisions to accelerate the growth of IP systems and "respond to pressure to be more competitive" (Dutfield and Suthersanen 2008, 22).

And indeed, since about 2000, developing nations have begun investing in their IP systems with new intensity, reworking them to not only coordinate their regimes with those of Western nations but also to pursue new titlings on an expansive range of living and cultural forms. Encouraged by international governing bodies—from the UN, to WIPO, to the World Bank—who advocate that developing nations exploit the IP potential of their national resources, countries have sought out titlings on artifacts that span everything from cultural and indigenous expression to biological diversity and local plant life (Berlin and Berlin 2003; Brush 1996, 1999; Castree 2003; Coombe 1998b; Correa 2003; Dove 1996; Dutfield 2002a, 2002b; Faye 2004; Gervais 2003; Hayden 2003; Isaac and Kerr 2004; Merson 2000; Nigh 2002; Parry 2000, 2002; Prakash 1999; Shiva 1996, 2004; Zerda-Sarmiento and Forero-Pineda 2002). These pursuits allow the expanse of a nation's cultural and biological diversity to be reenvisioned as not just protectable elements of national patrimony but also as exploitable national and economic resources. And they do so while offering the possibility of advancing culturally sensitive development models that are at once tailored to the demands of the twenty-first century's information-based economy. Curiously then, it is those nations with large rural populations, indigenous cultural traditions, and biodiverse territories that allegedly have the most to gain from such IP-based development initiatives. And as the case of Peru demonstrates, IP emerges as especially useful for development, not only because of the way in which it can be uniquely applied to the rural communities that are guardians of traditional culture but also for the way in which it invites and enables new productivities to be extracted from rural and indigenous subjects. It's these subjects who can now be figured into the nation's circuits of economic productivity and who can newly emerge as exceptional participants of the contemporary information economy—and indeed, even model actors in global enterprise.

It is precisely this notion of IP's productive capacity that's generated an almost tangible excitement coursing through the packed auditorium in Lima's National Museum that I find myself in one afternoon in July 2007, where the collective mood is buoyant. The audience has come to participate in the government- and United Nations–sponsored conference, "Folk Art, Innovation, and Sustainable Development," and we are now listening to Peru's national director of folk art (artesania), Madeleine Burns, speak on Chulucanas's new DO and the development of Peru's "national culture industry." Given the growing public appeal of such IP-based development strategies, it is not surprising that Burns's audience has flown in from around the world to participate in the conference or to hear about developments in Chulucanas's village, hundreds of miles away. In the first few rows of the auditorium alone are suited delegates from the United Nations, representatives from the Chinese and Indian governments, European design consultants, as well as representatives from Peru's government offices. With their laptops perched neatly on their knees, they listen attentively as Burns speaks, conjuring the remote, dusty Peruvian town hundreds of miles outside of Lima for her global listeners.

One of the first steps, she indicates, for the success of these initiatives is the making of a "native product." She asks, "What are the requirements to make a native product? First, that *artisans themselves* are the ones [who] decide what is their 'native product.'" That product, Burns specifies, should use local, regional materials and integrate "ancestral" techniques in its elaboration. It should also already be associated with a group of artisans who are committed to promoting its entry into and circulation in local, national, and international markets. "Native products," once properly realized, should help to not only create "value in local regions," but they should also further help to "define the identities" of their consumers and producers. The results of this kind of reform process, she emphasizes, are already noticeable with Chulucanas's IP title: "There hasn't only been an improvement of product quality because of the Denomination of Origin [in Chulucanas], it's also generated an interest among everyone [involved in ceramics production] to *improve themselves. . . .* It's generated a synergy between the public and private sector and the many institutions that have been involved with this issue—this development issue, that's ultimately about the generation of new capital."

Burns speaks passionately about the IP-focused work her office has undertaken in Chulucanas, shifting easily between the identity- and the market-oriented elements of the government's policy. It's not hard to see how she has gained the wide confidence of the teams of workers she has directed in her four years at MINCETUR. Her formulation of the phrase "the making of a native product" and what she stresses as its "identity defining" potential, however, are striking for precisely their apparent self-contradiction. What, after all, did it mean to make a "native product"? Shouldn't native artifacts—by definition—already exist as present in cultural expression, particularly if they were recognized as worthy of IP protection and distinction? Shouldn't the identities of indigenous producers likewise defy invention? Wouldn't the possibility of "authoring" native products—or their inventors for that matter—undercut claims to legal and cultural authenticity that IP titles mean to signal? And even if such inventive acts were possible, what stake did the state have in such authoring functions? Or in projecting such commitments for an international public similar to the one before her?

Curious as the notion of "making native products" may sound, it's precisely this work that has been channeled through Chulucanas and the Peruvian state's IP promotion initiative there. If Chulucanas ceramics are now newly distinguished and visible as productive sites of cultural expression, it is because they brought together a dispersed network of public and private forces all bent on remaking them as such. Their work was not only celebrated for engineering native products but also for, as Burns specifies, "generating interest" across a range of global and local actors whose collective activity allowed such artifacts to be mobilized for an international public. Distinct from other traditional crafts in Peru, Chulucanas ceramics and native products like them were intended to be disanchored from their local sites of origin and endowed with an unparalleled form of circulatory license. Such capacities, however, were not simply naturally present or effortlessly extractable features of native artifacts but turned out to have demanded new investments of work, energy, and resources from various actors in order to produce. Whatever the implications of its name, then, Chulucanas native products stand out emblematically as a unique invention, born out of the dynamic orchestrations of the Peruvian state and its network of transnational partners and refit for the information age.

Enterprising Villages

Critical scholars of neoliberalism have argued that the intensified interweaving between local sites and global publics are a key part of what characterizes contemporary conditions of economic globalization (Appadurai 1996, 2001; Canclini 2001; Castells 1996; Harvey 2006; Ong 1999; Ong and Collier 2004; Sassen 1991, 2002, 2006). It's the adoption of strategies that not only accommodate such interconnectivities but that also simultaneously favor the attraction of global capital and foreign investment that scholars emphasize as central to contemporary neoliberal policies (Harvey 2005, 2006; Ong 1999, 2005, 2006; Ong and Collier 2004; Sassen 2006). In efforts to encourage national economic growth and the accumulation of capital, governments actively seek means to liberalize regional markets, maximize international exchange, and reduce barriers to financial flows across national borders (Harvey 2005, 2006; Sassen 2006). In the process, government interests turn outward, decentering themselves from the realm of the national and realigning themselves to privilege activity and concerns outside traditionally national domains. State matters, that is, are reconfigured to foster the conditions that facilitate the entry of transnational interests into national infrastructure and public institutions and to render national economies permeable to new global investments (Sassen 2006).

Recent scholarship tracing these trends has stressed how globalization hasn't led to a disintegration of the state as much as to its redefinition (Appadurai 2001; Castells 1996; Harvey 2006; Ong 1999, 2006; Ong and Collier 2005; Sassen 2002, 2006). Such work has demonstrated how states' active participation in the processes of economic globalization is manifested, in fact, in a wide range of newly generated domestic work. This includes realigning national laws, legal norms, and legislation with foreign standards promoted by supranational organizations and leading industrialized economies to internationalize domestic markets. (Peru's own recent record of drastic legal reformations—captured but not encapsulated by the Fujimori administration's "shock" treatments and privatization policies—are not only reflective of these developments but are also illustrative of some of the boldest regional models of economic liberalization undertaken in Latin America.) Such work often entails new creative work, designing and erecting new infrastructures and introducing new "legalities" to accommodate and invite financial commitments (Sassen 2006). These new productions are undertaken with the dual intention of attracting transnational capital and formalizing the security of international corporate firms within national domains (Ong 2006).

But states do not only devote new energies toward the remaking of legal systems as an outgrowth of contemporary market-oriented strategies. They also dedicate efforts to accommodate the development of new urbanizations and hypermodernized zones as exceptional sites of economic productivity and centers of coordination for global financial transactions (Abrahamson 2004; Brenner and Keil 2006; Hackworth 2006; Massey 2007; Ong 2006; Sassen 1991, 2006; Scott 2001; Simmonds and Hack 2000; Taylor 2003). These so-called global cities serve to concentrate global capital and attract new international investments, channeling such resources toward the creation of spectacular, hypermodernized spatial arrangements, architectures, and forms. Scholars of contemporary operations of globalization have attended to precisely the emergences of these new postindustrial urban geographies and sites that encompass enterprise zones and technology triangles (Abrahamson 2004; Ong 1999, 2005, 2006; Ong and Collier 2004; Taylor 2003).

Indeed, such work underscored how part of the central functions of the neoliberal state is precisely the sustained production, attraction, and care of a diverse range of competitive knowledge classes and specialized experts that are attractive to global capital (Ong 1999; Zaloom 2010). It's these cosmopolitan subjects—skilled in the techniques of finance, legal, and technological management and creative production—who are responsible for working with the economic exchanges that traverse the networks of global capital (Sassen 1991; Ong 1999, 2005). Deploying new information and networked technologies to coordinate market-enabled transactions across space and time, these urban, knowledge professionals often become internationally mobile bodies themselves in the work of securing the constant, even flow of finance (Ong, 1999, 2005, 2006). Valued precisely for their specialized forms of expertise and optimized productivity, they are afforded an emergent form of transnational entitlement that not only permits an ease of international mobility but that also amplifies their ability to seek legal rights and exercise flexible, "citizenlike" claims in a diversified range of the sites. Unanchored to any single locale and traveling with passports and legal documents that authorize unhampered migration, these self-enterprising flexible citizens shared a cosmopolitan culture and economy of consumption across urban sites as disparate as London, Tokyo, Santiago, and New York (Abrahamson, 2004; Ong, 1999, 2006; Sassen, 1991). It's in such dispersed spaces that the tastes, lifestyles, and imaginaries of high-modernity's knowledge classes are trafficked and pool into one glossy, fluid blend.

The contemporary work of the Peruvian state in Chulucanas and its artisans, however, demonstrates how the unfolding of these trends has not left rural spaces or local, rural populations disconnected. In the market's quest to optimize production and uncover new commodities, marginal spaces and traditional populations selectively acquire new relevance. Such relevance is based not only on the possibility of unique and potentially nonindustrialized techniques of production but also on the market distinctiveness acquired from a product's ties to the traditional communities and remote or rural localities outside the circuits of global cities and their professional classes. Parallel with emergent global cities, where the productive labor of urban knowledge workers and shared patterns of consumption can be concentrated, arise new rural satellites—or what can be called *enterprise villages*. In these remote spaces, unique forms of production can be made hyper-efficient. Such productive efficiencies are extracted not merely along the criteria of time scales and cost of manufacture but also in the production of diverse commodities that range from indigenous ceramics to computer chips and that capture the market potential in traditional knowledge and high modernity. New tools and resources of the information economy—including legal instruments such as IP titlings—have played a key role in connecting rural sites to global channels of exchange and in allowing such zones to distinguish themselves, among other sites and potential competitors, as optimal network partners.

If it is new territorial alignments that globalization creates, then it does so by more than simply dividing space into uniform geographies of urban investment and rural divestment or classifying populations into categories of urban economic privilege and rural, remote abjection. Knowledge workers from global cities emerge as active, model participants in the workings of global capital but so, too, are rural subjects being called on to participate in optimizing the productive spaces of the global market. If it is a lifestyle lived within a circuit of hypermodernized sites that urban knowledge classes cultivate, rural subject participation allows such a network to extend into and occupy new marginal spaces. And through these resettlements, rural and indigenous subjects may now emerge as a unique breed of knowledge worker. Incorporated into global capital's network as guardians of traditional knowledge, rural artisans invest their own creative labor into global markets. And they can do so as a specialized class of workers whose productive energies are authorized and enabled—not only by the possession and use of passports and other resources of the modern state (Ong, 1999) but also by the possession and use of artisanal tools as simple as the wooden paddle.

Indeed, that rural and indigenous subjects may now become newly incorporated as not just potential contributors to the global economy but as model participants of it is a celebrated achievement of the Peruvian state. It's precisely this consciousness of artisans as potential creative laborers that can stretch the margins of who may be included in the knowledge economy that informs a concerted emphasis on artisans themselves in the state's development policy in Chulucanas. There has been minimal labor devoted, then, to reforming conditions external to Chulucanas that affect the consumption and circulation of artisan products. Little visible work has been done to attempt to penalize or diminish the incidents of either local or nondomestic "piracy" of Chulucanas ceramics (and to reduce the number of copies sold in shops in Lima, Ecuador, or China) or to educate consumers and exporters alike on the historical and cultural roots that distinguish Chulucanas ceramics. Efforts have instead focused on "training" rural artisans themselves in the techniques of international marketing and entrepreneurship and reengineering the workshops of artisans into efficient production zones. Since the work of developing a DO for Chulucanas began, MINCETUR and INDECOPI have organized routine classes and conferences for Chulucanas artisans that feature international exporters, market consultants, and engineering and design professionals from Lima. The idea has been to reskill ceramics makers so that they behave less like provincial artisans and more like sophisticated, market-savvy entrepreneurs and IP rights holders who can respect and navigate the logics of the global market. Or as Burns herself later tells me, "I believe what we are all subscribing to is the issue of competitiveness. And the state's responsibility in this respect is what it has developed through workshops, legal marks, providing tools and infrastructure. This is a union of policies that are directed towards improving the competitiveness and utility of the producer, who is part of this area of [national] growth."

The newly shifting awareness of the productive potential in traditional populations and indigenous subjects has likewise been reflected in the institutional adjustments of INDECOPI itself. Established under the neoliberal reforms of Alberto Fujimori's government during the 1990s, INDECOPI's mission was to strengthen private property rights and coordinate Peru's IP and patent system with those of leading industrialized economies. Considered by many government officials as among its most effective public offices, INDECOPI's IP-oriented reforms are seen to have secured, protected, and increased foreign and domestic private investments over the past two decades. Until recently, however, INDECOPI saw itself as bearing little relevance for the majority of Peru's population living outside of cities or to indigenous populations scattered throughout the country who historically had been disconnected from the state.

The recent development of IP titles for rural communities such as Chulucanas marks a pointed policy shift for INDECOPI to innovate and offer up legal tools for Peru's nonurban and indigenous populations. In the last several years, INDECOPI established two new offices that are meant to address the rural and traditional communities who are now newly envisioned as relevant productive classes. Empowered with such tools, it's believed, the productive potential of such classes can be readily extracted. Among the newly instituted directives of the office, then, is a goal to establish two new IP titlings as DOs for traditional artisanal and agricultural products every year. Indeed, although only two DOs existed in Peru until 2005—that for the brandy Pisco and another for Cuzco's maiz gigante—since then, the number of active DOs have quadrupled. Following Chulucanas's DO, new titles have been issued for coffee from Villa Rica and Huadquiña, respectively, pallares beans from Ica, loche pumpkin from Lambayeque, maca tubers from Junin, and cherimoya fruit from Cumbe (with five more in early stages of processing as of this writing). Officials from Lima also recently began traveling to remote rural zones to visit indigenous communities and organize workshops that encourage them to begin to create inventories and archives of traditional herbs, plants, and medicines from which patentable material may be extractable.

But if the Peruvian state has begun to expand its focus from urban zones and subjects to the very rural sites and traditional communities that had been historically neglected, it's crucial that it has begun to do so armed with what it sees as the transformative potential of IP. Such transformative capacities can be channeled through these information resources in several ways. First, the state's use of IP inserts *culture* into a logic of information property that assigns it a dual valence. Although IP here recognizes culture's inherent value, it also calls for it to be transformed in order for its market potential to be extracted. Second, IP inserts culture into a global web of relations and infrastructure to sustain these optimizations. These include Peruvian ministries, supernational governing bodies, transnational corporations, and international NGOs. And third, IP allows a selective focus not on the entire body of citizens who share an interest in culture but only on those subjects deemed legitimate: "IP rights holders." Investments, then, go only into those enterprising individuals who are most able to network IP's productive relations.

It has been precisely this capacity to optimize market potential that gets underscored in Madeleine Burns's emphasis on "improving" and creating "new capital" in Chulucanas. Significantly, if IP titles are being applied to Chulucanas ceramics now, it has been with a logic that the work of optimizing culture can never actually be done. By global capital's clock, new production efficiencies and profit margins must continually be extracted—and IP is a tool in this new work. What Burns does not say is that there may not actually exist an absolutely "perfected" good. Traditional goods—like the rural zones from which they originate—can only be in the process of improvement, ever-relegated as imperfect, and subject to reform to meet shifting market demands.

What circulates as one of the primary justifications for the cultural application of IP titles—that is, that an unpolluted, indigenous culture or a virgin, natural territory can receive public protection and recognition for its "native" purity under such legal titles—is undone here. Nothing, perhaps, evidences this more than the exclusionary function of Chulucanas's DO and the fact that the vast majority of Chulucanas's artisans will not have the right to exercise the IP title or be a beneficiary of its rewards. Of the more than four hundred artisans with workshops in the town, only six have been granted the right to use the IP title to date. Strange as this may seem, local developments in Chulucanas suggest that it is in fact an intended effect of the state's electronic government initiatives, whereby disparate actors—including rural artisans, Lima-based state representatives, and global exporters—invested in Chulucanas's ceramics market are brought into selective alliance. The state assumes the work of cultivating *flexible relations* (or what artisans narrate as *relations of promiscuity*, as we'll see in chapter 3) between actors with diverse and often oppositional interests. It's the new linkages among these actors—rather than one's generational or familial ties to ceramics making or ties to environment and resources that are part of that creation—that now determine how benefits, rewards, and protections will be channeled.

It is not merely the opportunity to reap new economic benefits, however, that compels these networked couplings and that seduces disparate actors into alliance. It is as much the promise of "global connection"— and the opportunity for diversely situated subjects to expand a sense of individual agency and realize themselves as distinctive, globally enabled competitors that sustain such networks. They are inspired by what political theorists, Andrew Barry, Thomas Osborne, and Nikolas Rose (1996) describe as the promise of a "form of politics beyond the State . . . [that] emphasizes the crucial political value of the mobilization and shaping of individual capacities and conduct." Embedding oneself into networked relations that pronounce the ability to "independently" build linkages with authoritative bodies beyond the boundaries of the nation ironically becomes a means of gaining state recognition. But if the case of Chulucanas demonstrates anything, it is that there is no shortage of contradiction. Networks channel wealth as much as they deny it. Relations of efficiency and productivity on which new fortunes are based at once undermine preexisting relations of communal life and the productive relations within. Tradition, ancestral knowledge, and culture emerge not as revered, sacred objects worthy of protection but as objects that are instead supremely malleable, porous, flexible, and subject themselves to new permeations both material and informational. In a world where economic opportunity and security are experienced as scarcities, it appears that new forms of individual risk are voluntarily assumed as a means of attempting to minimize other social, economic ones. And however troubling these conditions, the final irony is that the success stories still abound.

Fortune's Network

Although the state's IP-based initiative in Chulucanas is only several years old, the work of turning Chulucanas ceramics into an exportable "native" product in fact began more than a decade ago as part of a large, $40 million USD development project funded by the US Agency for International Development (USAID), the branch of the US government that finances global development programs. The project sought to revitalize Peru's rural economies and reestablish foreign consumers' trust in local markets in the aftermath of Peru's more than two-decade-long civil war with the peasant-based Maoist Sendero Luminoso (Shining Path) movement (and USAID's funding continues today to "advance US national security, foreign policy and the War on Terrorism"). By cultivating export markets for artisans, it hoped to rebuild communities that had experienced decades of unrest and economic disruption during the civil war as well as undertake measures to prevent Sendero's resurgence.[1] To deploy the project locally, USAID partnered in Peru with the Peruvian Export Promotion Agency (PROMPEX), the Association of Exporters (ADEX), and the US-based nonprofit, Aid to Artisans (ATA).

When speaking before her audience, Burns's emphasis on IP's ability to generate "new capital" while synergistically "improving" artisans and products is crucial precisely because such new IP investments in Chulucanas actually seek to further optimize the array of earlier international investments there. It's no stretch to say that the state's new IP-oriented investments in Chulucanas were awarded based on its proven record of economic production, and that its present granting comes not only with the recognition of past performance but with expectations of future heightened productivity.

Indeed, for the institutions that were involved with cultivating Chulucanas's export market, the case serves as an emblematic example of the success of their efforts. And in project partners' retellings of the development of Chulucanas, there is no shortage of talk of the salvational—almost transcendental—power of the global market there. Participants readily tell me that by the end of the project, thousands of new jobs were created and craft exports more than doubled nationally to $23 million USD, the largest portion of which came from sales of Chulucanas ceramics. Until the 1970s, Chulucanas ceramics were relatively rare objects to be found in a handful of boutiques in Lima, Europe, and the United States, where they fetched prices upward of $150 USD. Or for the determined consumer, they could have also been found in a few workshops in Chulucanas, where pieces—whose traditional, handmade production allowed no more than a dozen to be produced in a week—could be purchased for a third of what they sold for in urban and foreign boutiques. The USAID, ADEX, and ATA collaboration transformed all that, "modernizing" traditional production techniques and spaces, "optimizing" workshops and their output, vastly increasing global distribution, and broadening the base for international consumption.

By the late 1990s, Ceramicas Chulucanas was no longer the same artifact. It could be found easily in Lima's budget-friendly tourist markets and high-end souvenir shops. US chains from Neiman Marcus and Pier One to Target and Ten Thousand Villages sold Ceramicas Chulucanas. Increasingly visible as a good that attracted foreign capital, mimicked "pirated" versions of it began to circulate—first in regional markets, then in Lima and beyond. Such happenings affected Chulucanas work patterns, too, where the number of workshops and skilled workers began to explode. In the beginning of the 1980s, no more than a dozen artisans were represented under Chulucanas's oldest artisan association, the Association of Vicús Ceramicists. Today, more than two hundred artisans are members. More telling, perhaps, is that the Vicús Association now exists as just one among five other ceramicist associations, which at times see themselves as competing for representative status of Chulucanas ceramicists. In just a few decades, Ceramicas Chulucanas had gone from relative public obscurity to becoming today one of Peru's most recognizable folk arts and rural exports, so much so that in 2005, the government honored it by naming it a "national product" (Producto Bandera), the only folk art among Peru's hundreds of different craft traditions to be distinguished with such a title.

Sonia Cespedes was the sociologist and development consultant contracted by ADEX in the mid-1990s to complete a technical assessment of production conditions. She was among the first of the team of professionals working as part of the Chulucanas development project who arrived there in 1995 to work with the artisans. She spent the next several years of her life shuttling between Lima and Chulucanas, laying down the groundwork for what she calls Chulucanas's technological revolution and working alongside other economic and design consultants and the artisans from Chulucanas. I first spoke to her in her home office in Lima's middle-class Surco district, where she has also built an extension for her own ceramics studio and workshop. The walls of her home are lined with examples of her own ceramics work and one notes in several pieces the influence of the negative-positive design technique, one of the signature traits of Chulucanas pottery. She hasn't traveled to Chulucanas in several years but she speaks fondly of the collaborations she helped to engineer there, the lifelong relationships that sprang from them, and a history that she tells me was personally life changing. She asks during our conversation about several artisans by name, just as she asks to hear about the news and changes in particular sites and locales there. And I learn later that she is the godmother of the children of one of Chulucanas's most prominent artisans.

She tells me, "This was one of the most important experiences of my professional life. This kind of fortune is something that one doesn't always find in their professional life, this fortune of finding ourselves as the right people in each link in the chain." She lists for me more than a dozen names of USAID personnel, Lima-based exporters, US-based designers, and Chulucanas artisans whom she worked with, who became part of a "chain that carried us to freedom." She specifies that in order for the network to have functioned as it did, each participant needed to understand what part, what circumscribed identity, he or she was scripted to play. She stresses this for me to make sure it's clear: "Each key person had [his or her] role, and each had [his or her] place. So it was this confluence between all these professions and professional roles through which this fortune emerged."

A number of key objectives were identified by the project, including making the necessary "improvements" in the quality, design, and price of Chulucanas ceramics for them to become exportable. She recalls, however, that it was no small task to first convince Chulucanas artisans to participate at all in the export-oriented partnership or to assume the roles such a collaborative model prescribed for them. "I tell you that really, it was tough work (*trabajaba como una hormiga*).[2] A lot of effort was required. I had to go door to door, knocking on each, speaking to each workshop and each artisan about the benefits that this project could bring." The majority of the artisans she spoke to responded with either disinterest or with a pointed rejection of her invitation. A small group of artisans—no more than four she recalls—did, despite the resistance from the majority of their fellow artisans, finally decide to cooperate with the project: "This small group had a lot of faith, and it was this small group that became the force behind Chulucanas. . . . Of course there were many in the beginning who didn't agree, but now no. Now it's another Chulucanas. . . . The exporters said to them, 'Chulucanas, improve your designs, your production, your quality, your price, and we'll commit too!' That's why I tell you that each link in the chain believed firmly. And we launched ourselves into this marvelous adventure! [An adventure] not because we didn't know where we were going, but because we did and we had the nerve to believe it was going to work!"

Spurned in Chulucanas for their reform initiatives, the export-oriented team could read themselves as "daring" and "rebellious" precisely because they challenged long-honored traditions in ceramics making and the social conditions that sought to conserve them. If the pro-export reformers provoked the scorn of the town, then, it was in large part because their project sought to introduce radical transformations that threatened to destabilize the social world of ceramics making around which community, customs, and a history of collective belief had been formed. But the language of rebellion that exporters adopt presumes a courageous "break" and "rebellion from" a shared space of conserved tradition. It is a language, in other words, that situates them alongside the artisans of Chulucanas as if they had been bound together by a shared, past alliance. That such identifications never actually existed is overlooked by exporters—eclipsed instead by a vision of a boldly remade future that their reform proposals project. And if subscribing to the pro-export reform agenda demanded a kind of "faith" from coupling artisans that even exporters could acknowledge, it is telling because it would be these artisans whose rebellion from established community norms would be required before they could partner with the band of unknown strangers whose calls for change had upset the town.

One of the most controversial changes the partnership demanded was that the artisans radically lower their prices by a factor of four, drastically reducing the cost of a ceramics piece that Cespedes tells me would originally be priced at an "unthinkable" $30 USD. Prices were adjusted to $7 USD for the same ceramic, a calculation derived after an analysis of the cost and abundance of the raw materials needed for production. The clay, whose source was local earth, sold cheaply and was in plentiful supply. And the sheer evidence of this could be used to persuade artisans that previous pay scales for labor was in fact overvalued, enabling the generation of a new supply of unpaid labor. Cespedes recalls, "I never imposed anything on the artisans. I said to them, 'look, at this [economic] study that I'm completing with you. We've gone to the clay pits and bought clay together, and we've seen that the prices [you're asking for] are too high. You are tripling prices of a product that's not of quality. Why don't we reverse this model and first lower the cost of ceramics by a third. . . . You have to be conscious of how much the exporter is going to gain. . . . You have to know how much the percentage of profit is so that you can learn to negotiate and know until how much you can raise and lower your prices. . . . Because this isn't about if I like something or not. Everything has a technical explanation, a mathematical explanation. These things aren't done with the heart, but *with reason.*"

If Cespedes had to devote significant energy to translating exporters' pricing matrix, it was because it indeed sought to challenge the common sense of artisans that was rooted in local life and traditional production. Pricing, by the exporters' logic, was not to be determined with a "heart-based" attachment to artisanal tradition or sentimental overvaluing of local culture among them but with reason-based, technical assessments on the external dispersion of profit. According to such an analysis, the market potential of Chulucanas ceramics had been "irrationally" inhibited by local traditions and sentiments of production. To ensure the efficient distribution of local products and the dispersion of profit beyond the artisans of Chulu-canas to an array of actors spanning across the globe, new assessments and considerations of expanded forms of nonlocal work were required. Such analyses effectively revalued artisans' local labor by determining what pre-vious forms of regional craft and artisanal production were being errone-ously overpaid, and newly recognizing distributed forms of nonlocal work as necessarily underpaid. It would be in these new extensions of nonlocal labor that profit around Chulucanas ceramics could be captured and on which the realization of its "market potential" was to be based.

Cespedes devoted much of her work to running workshops that sensi-tized artisans to precisely these logics of the global market and the means by which commodity prices and profit margins were to be deduced. "They had to learn how to think in this way. . . . [So I said to them], let me impart this to you so that you can use it, rather than protest it, because it's going to help you." She also organized the artisans into teams that would work under each of the three primary export companies—ALLPA, American Trading, and Berrocal, Ltd.—in Chulucanas. She remembers vividly how her suggestions initially made her the object of revile among Chulucanas artisans. She insists that to progress, however, one has to have an "open mind, professionalism, conviction in your product," and above all, the "nerve" to adopt radical change, even against popular will.

But if it's a narrative of courageous change that pro-export actors such as Cespedes adopt, it's crucial that by the register of Chulucanas artisans, their reform work could be described as not brave but brutal, not only because their export-oriented reforms insisted on revaluing of traditional techniques as an error in order to cultivate new market extensions but also because such reforms entailed diminishing the actual value of local labor. And if export-oriented reformers encountered an initial resistance from the Chulucanas community members, who were almost uniformly opposed to their entry, it was because their work implicated local craft making and forms of rural production as "overvalued." By reformers' accounts, these were market inefficiencies that they intended to correct. Such work could only be sustained, however, with local collaboration. And Cespedes, who still consults on rural development projects and travels across Central and South America to do so, attests that she did secure these local commit-ments. "In the end, my trainings became so important that first, everyone [working in ceramics] wanted to get trained! Exporters only wanted to work with artisans that we had trained, and even asked for their certifications that we had trained them."

But the popularity of Cespedes's workshops and their pro-market preachings eventually gained underscores a curiosity of neoliberal logics and the mechanisms by which they recruit new participants. If reformers faced an almost uniform resistance of community members on their initial arrival, the events that unfolded in Chulucanas illustrate that combatting such opposition—and converting opposers into allies—and thereby securing continued diffusion of export logics was achievable. If there was an audacity to the development of the Chulucanas export market, then, it is perhaps that it became possible at all to undermine the staunch resistance and scorn that reformers initially faced. In the process, reformers themselves could be discursively transformed. Those who insisted on the overvaluation of local culture as a market error could later posit themselves as defenders of Chulucanas's possibilities of new future wealth. And even if such new profit would be largely accumulated by parties outside Chulucanas (or was only directed to the discrete group of artisans who first "risked" coupling with exporters), reformers could still speak in populist tongues, insisting that the possibility of acquiring wealth was within reach of any local resident—given the capacity to grasp market logics and a willingness to think outside "one's heart."

If pro-export preachers could insist on their rebellious nature, then it rested in part on their ability to have bridged the tensions in their gospel and on their capacity to convert their strongest critics into willful partners. In the process, export-oriented reformers can be reborn, too, as some of the most effusive defenders of and spokespeople for village life and local culture. Cespedes, notably, spares no emotion when she stresses that Chulucanas culture is more alive today than ever. In protective tones, she tells me, "Today, Chulucanas looks as if it were a city of all artisans. . . . The culture Vicús isn't going to die. It's something transcendental." And I can't help but want to believe her.

A Networked Conscience

On a sunny, summer afternoon in November 2006, Javier Escandón is speaking to a roomful of artisans in the offices of the MINCETUR-managed Center for Technological Innovation (CITE) in Chulucanas. His tall frame casts a long shadow against the blue glow of the PowerPoint projection behind him. His presentation is intended to prepare the artisans he's speaking to for the upcoming international craft expo in Lima, The Peruvian Gift Show. The expo, organized jointly by MINCETUR and PROMPEX is *the* national event of the year for folk art sales. It attracts thousands of foreign buyers who come in search of local Peruvian crafts and their suppliers. These buyers come with the mission to choose who among the hundreds of artisans and suppliers available to them—each with their display stands positioned alongside each other—they will ultimately select to contract with. These couplings for the buyer mean stocked shelves and sales rooms for consumers back in Europe and the United States. For artisans, it means months of steady labor for themselves and the team of workers employed. Access to these fairs as one of the primary spaces that artisans can use to secure contracts and build new relationships with international buyers is highly coveted. Because vendors can set up their stands only by invitation, however, just two of the artisans in the room will actually travel to Lima for it. The two that will travel, among the handful of artisans in all of Chulucanas who have an established track record for contracting and completing large export orders, have in fact traveled to the fair before. Still, the two dozen or so other artisans in the room who have turned up for Escandón's presentation know that his advice will be of value to them nonetheless. Escandón's own work history in Chulucanas, after all, began more than a decade ago when he was the director of the ADEX and Aid to Artisans project there, and similar to Cespedes, he maintains his links to many artisans working there. And importantly, his ties to exporters in Lima, in particular, the three largest exporters of Chulucanas ceramics, still run deep.

The slide that Escandón projects behind him displays the image of a modern living room of a nonspecific end consumer's middle-class home. With the text "life, peace, tranquility, security, and naturalness" typed beneath it, the home displayed could be any in Lima or New York or London. For Escandón and the artisans his workshop is tailored for, the geographic specificity matters very little. His purpose is rather to project an image of the global consumer, to convey the contemporary consumer's values and identity captured in the very domestic interior projected on the slide, and to orient the market consciousness of local artisans accordingly. He tells them, "What one seeks is a sensation of peace, of tranquility, especially because it's such a violent world that we live in now. . . . And when you look at the product, it shouldn't be mistaken for something that came from the US or Europe. [It should] capture *our* identity . . . [and] still be recognized even on the other side of the world, that it starts a global dialogue."

One way of reading what he means is that artisans should see themselves as responsible for engineering "global dialogues" with consumers who are

culturally diverse and globally dispersed. And he means to stress that if artisans intend to court the global market, their "globally dialogic" products
have to be traditional and modern at once. They must be able to deliver a
traditional, cultural good and be ever-conscious of the "global dialogue"
that the product is now meant to operate within. And they must do this
while still meeting the scaled-up production demands and "high" quality
expectations of the international market. They must, in other words, be able
to split their consciousness between the mixed anxieties and desires that
flow through the local world of Chulucanas, those of the foreign importer
navigating the Lima market in search of exportable goods and those of the
anonymous end consumer half-way around the world.

He leaves them with one final piece of crucial advice: "When you go to
an international fair, the last thing you want to do is mistake it for a tourist
market. . . . We've seen that these fairs require export vision and capacity,
knowledge of each of the products to be shipped, product prices that are
already fixed according to the market analysis. . . . Because when these
buyers come, they aren't here to tour. And this is very important to keep in
mind because it's what will transform you into someone who can relate to
an importer, what will classify you as a 'person of potential' with whom it'd
be possible to build a sustainable relationship over time."

Listening to Escandón's presentation and his appeals to artisans to consider the "global dialogue" their artifacts should spur and the "persons of
potential" they are meant to become by participating in the international
market, it's easy to imagine the address as part of a university lecture on
global consumption. If this is the work of twenty-first-century development
and reform that's being undertaken in these skills-building workshops, it
is certainly not with the same patronizing, Enlightenment mission that
characterized nineteenth -century schools as projects of the modern, liberal state (Alonso 2005; Barry, Osborne, and Rose 1996; Joseph and Nugent
1994; Mallon 1994; Rockwell 1994; Rose 1999; Wilson 2001). Under the
worldview of the new, ruling elite of the young Peruvian state, the country's remotely situated, rural, and indigenous populations were read as
"ignorant, defective, and uncivilized" (Wilson 2001, 326). Such populations could be incorporated into the nation as modern citizens but only
after being transformed under Peru's first national education project. José
Carlos Mariátegui, one of Latin America's most renowned socialists and
political philosophers who worked as a journalist in Peru during the early
twentieth century, had pointed to the explicitly "colonial" nature that was
maintained by the national education system and critiqued the means by
which it regarded indigenous populations as "an inferior" race rather than

as "equal" citizens in his famed 1928 publication *Seven Interpretive Essays on Peruvian Reality* (1988). Escandón's speech, however, unlike the civilizing discourse Republican elites of the nineteenth and twentieth centuries adopted, doesn't condescend or pretend to uplift, enlighten, or save its rural listeners through reason. His language is riddled instead with a firm "you can do it!" ethic, with a steady faith that the productive individuals and "persons of potential" before him—however remotely located from the centers of financial transaction and its urban knowledge classes—can be responsible for improving their own conditions, and that they only need to be given the opportunity to freely compete in the market to demonstrate their ability to do so.

True to his faith, Escandón infuses markets here with the potential to deliver rural subjects and zones from an eternal condemnation to poverty, presenting global markets as an option "that if rightly harnessed, is invested with the capacity to wholly transform the universe of the marginalized and disempowered" (Comaroff and Comaroff 2001, 3). Critical scholars of neoliberalism have thickly described this before—how it is that free market advocates appeal to a salvational potential of millennial capitalism in preaching their gospel (Comaroff and Comaroff 2001, 2005). But there is a crucial difference here. Whereas millennial capitalism billed itself as saving subjects, here redemption now depends on the cultivation of subjects with the capacity to self-save. They are addressed by a neoliberal logic that notably "doesn't seek to govern through 'society' but through the regulated choices of individual citizens, now construed as subjects of choices and aspirations to self-actualization and self-fulfillment." Construed as "subjects of choices," individuals are governed under a logic that seeks to govern by "implant[ing] in citizens the aspirations to pursue their own civility, well-being and advancement" (Barry, Osborne, and Rose 1996, 40). But, crucially, it does so under the presumption that it addresses not flawed and ignorant savages but enterprising and proficient individuals with the capacity to achieve their own self-actualization and self-fulfillment. This is, in other words, a messianic capitalism tailored for the twenty-first century that doesn't need to come to the rescue—but that offers up to subjects the tools for auto-rescue. And that manages to seduce precisely in its promise to allow subjects to self-realize as free, modern, and optimized individuals.

Indeed, speaking to Escandón later in his home in Lima, he embeds Chulucanas history of development in a narrative of successful geographic competition: "Now you've proven that Chulucanas can do it. In no other part of Peru can you generate the [quantity] of export[s] that you see there. Just in Chulucanas. . . . The important thing is that with Chulucanas's productive organization, it proves that it has the capacity for production. It works *like a business card* that shows you can work in large volumes in short amounts of time. It's something that hasn't been done in any other part of Peru." His description of Chulucanas as having achieved a productivity still unmatched by any other remote region in Peru is a characterization I hear echoed repeatedly about the town and its unparalleled, optimized production power. I am meeting with Escandón in his home shortly after the Peruvian Gift Fair, which succeeded in generating various new sales contracts for the Chulucanas artisans who attended. Still, for Escandón, it was crucially flawed in at least one way: "The state was subsidizing the artisan so that he could come to the show. This to me doesn't seem right and I wrote this in my report [to ADEX]. . . . You are going against the law of market! . . . We're talking about what I would call unfair competition. And I don't agree with this kind of work. I don't think it helps the artisan."

In Escandón's living room, an ample space with high ceilings and even lighting, a number of artisanal pieces, collected over decades of work with artisans and exporters across the country, are tastefully displayed. Escandón is seated across from me, his long legs crossed leisurely so that one loafered foot is balanced inches from the ground. He never fails to impress me with the ease by which his body language conveys a natural, unshakeable elegance, even when something has upset him. Indeed, weeks after the Peruvian Gift Show's end, Escandón is still visibly bothered by the thought of the state subsidies that financed artisans' travel expenses to the event. It's as if it offended a deep moral ethic in him: "The artisan should be accustomed to paying for his own things, just like any one of us."

Authored Selves

Gaining entry into international crafts fairs, however, was just one key indicator that rural artisans had successfully cultivated their own competitiveness. Another indicator was the ability to acquire one of the much coveted contracts with the handful of companies that export in large quantities to international retail chains. Vying for these contracts and coupling with exporters has obligated workshops to undertake major transformations in order to meet production demands. And it has meant leaving behind traditional production techniques for modernized, large-scale production models in which newly incorporated technologies replace much of the work originally done by hand. Rather than shaping ceramics by paddle or by hand, modernized workshops now use electric potter's wheels and ceramics molds. And whereby ceramics making had once involved a single artisan completing all the production steps alone or with a few family members, modernized workshops now employ teams of workers who are each assigned a single, dedicated task.

Such transformations have amplified production potentials from the several dozen pieces traditional workshops produced to some three thousand ceramics or more monthly. But for artisans who are willing to commit to such self-modernization efforts, vast amounts of new work are generated. Labor must now be expended not only to ensure that the required changes be deployed properly but also to ensure that the established production models and the relations that sustained them (now to be displaced in the wake of new transformation efforts) are also properly handled. New export contracts come with new obligations to ensure that workshops are properly modernized and that new business relations are successfully negotiated. Managing this added labor at all, however, is more than a means to remaking space and evidences an artisan's capacity to perform as a self-realized, modern entrepreneur. As much as remade workshops bespeak of the investments expended to transform production spaces, they demonstrate an artisan's own investment in individualized self-transformation.

Among the artisans who are recognized in Chulucanas for having managed to successfully negotiate these new dual responsibilities is Antonio Lopez. Born into a family of traditional potters, the forty-year-old Lopez was one of the first in Chulucanas to integrate modern production technologies into his workshop, enabling him to produce some of Chulucanas's first large international export shipments. And although most of Chulucanas artisans have yet to attend a single fair in Lima, Lopez has already cultivated an extensive history of fair expositions, having attended several in Lima and having traveled to others in Brazil, Mexico, Finland, Columbia, Holland, Germany, and Chile. He remembers having been mentored by Celestino Vera, one of Chulucanas's most distinguished ceramicists, and one of the few who still makes his pieces only by hand. During the 1970s, when Chulucanas ceramics were still only made by hand and only beginning to find a distribution market in a few select boutiques nationally and internationally, Vera invited the younger Lopez to travel with him to various fairs he had been asked to attend: "He asked me to come, and I came with my own products, which sold more or less well, and that's how I began. . . . But Celestino wasn't my teacher. I learned alone, I was *self-taught*, because I created my own line of ceramics, with a design that was totally different—with plates and vases—which [sold well]."

Lopez repeats such assertions of self-made success built on personal achievements and hard-earned recognition throughout my conversation with him in his workshop, a large space where the flattened earth still serves as the floor. Similar to most of Chulucanas's ceramics workshops, Lopez's extends from his home. And often, it's a doorless threshold or hanging sheet that marks the end of an artisan's workshop space and the beginning of kitchens and bedrooms. His workshop is one of the few in Chulucanas, however, that has a separate office space (complete with a coveted cement floor) built in. As he speaks to me, he molds a ceramics piece using one of the electric potter's wheel that's now become customary in the production of exportable ceramics. His quick, fluid strokes flow like second nature from his arms over the wet clay, forming a base and neck of the vase in a matter of seconds. I notice that all the while, he manages to keep the pale, collared short-sleeve shirt he's wearing unmarked. When he speaks again, it's to acknowledge the generosity that Vera extended toward him during his early career. He recognizes how it played a crucial part in the course his life would later take but he doesn't dwell on this: "Celestino wasn't selfish, and that's why he invited me."

But such expressions of generosity and support among artisans in Chulucanas have become a rarity today. Lopez's explicit recognition of Vera as "not selfish" is the one of the only instances I witness when such kind words are expressed by one artisan for another. And it's one of the few instances when I've heard an artisan acknowledge having been the recipient of a kind act by a fellow artisan. More commonly what I hear when speaking to artisans are expressions of explicit distrust and heightened competition. Even after the state's work of promoting Chulucanas as the distinguished site of cultural expression—work that one might have assumed would foster a sense of collective pride and community—accusations of betrayal against other ceramicists, reports of design stealing, accounts of price warring, and rumors of wage exploitation abound in the town.

Josefa Nolte is an anthropologist who first began to work with Chulucanas artisans as the director of the crafts-oriented nonprofit organization Antisuyo. Nolte and Antisuyo were present in Chulucanas when the USAID-ADEX project was undertaking reforms there. Antisuyo's showroom in Lima were known for selling Chulucanas handmade ceramics with some of the highest prices, and Nolte defended the pricing scheme after the USAID-ADEX collaboration began to advocate its price adjustments. Artisans I spoke to recalled with sadness when in the late 1990s Antisuyo had to close its doors and stopped showcasing Chulucanas ceramics. ("I guess it just wasn't sustainable," Lopez told me.) Nolte's own sadness when explaining the developments in Chulucanas and the heightened competition now part of the landscape there is audible: "[Ceramics are] an economic resource in a place where there aren't many resources."

She is accusatory when she speaks about the companies who developed Chulucanas's export market, describing their relationship to the town in almost predatory terms. "The exporters all made their fortune in this period. . . . [T]hey got strong and large in this period, through their export sales." What offends her most, however, is the lack of commitment the exporters demonstrate for the artisans whose labor they depend on. "The largest exporters now aren't interested in working with small artisans. They're only interested in whoever can offer them the largest volume, the best investment, those who have already grown. And this bothers me, it angers me. . . . Maybe tomorrow I'll decide that I don't like selling crafts anymore, even though I built my fortune on this. Tomorrow I'll leave and I'll build a hotel. Or I'll sell something else, it doesn't matter. I've made my fortune, I've grown, and now I don't need you. Ciao." The hyper-individualism, however, and lack of commitment to community is something she laments is now notable among artisans as well: "We have to work hard with the population in Chulucanas, because if the *population doesn't have any conscience*, if they are not united, they are never going to achieve anything. Because if you tell me you will sell [a piece] to me for four [dollars], I will go to the guy next door and he'll tell me he'll sell it to me for three. Who benefits? The dealer in the long and in the short run."

Nolte's critique of Chulucanas's export model—and her concern for how it's fragmented the population of Chulucanas into those with and those lacking fortune—echoes an observation critics of neoliberalism have made before. Alongside the rapid assembling in global cities of hyperproductive zones that concentrate and accommodate wealth and comfort among "the fortunate" have arisen new—often neighboring—sectors of economic need, disregard, and abandonment where those categorized as un- or underproductive are concentrated (Atkinson and Bridge 2005; Biehl 2005; Klein 2007; Lees, Slater, and Wyly 2007; Sassen 1991). Populations have been divided between "individuals who possess human capital . . . [and those] who are judged not to have such tradable competence or potential" (Ong 2006, 7). Applying a "logic of exception" (Ong 2006) stratifies populations along the criteria of competitivity and marketability, marking out which subjects are worthy of exclusion and which are to be selected as deserving targets of official investment, care, and protection. Esteemed as value-creating individuals, self-enterprising subjects are permitted expanded social entitlements that allow them to make not only "citizenlike benefits" in new nonlocal contexts but that also afford them a kind of spatial mobility as delocalized "flexible citizens" (Ong 1999). In contrast to these hyper-intensified modalities of living, however, arise zones of abandonment where the "social death" of the biologically living but unproductive finds a dedicated site. In such zones, "one is faced with a human condition in which voice can no longer become action . . . [where] the human being is . . . knowing that no one will respond, that nothing will open the future . . . [and where] absence is the most pressing and concrete thing" (Biehl 2005, 11). Much as the logic of flexible citizenship presumes that their intensified entitlements are earned privileges built on a record of achievement and competitive individualism, so, too, do logics of abandonment presume that the forgotten and left behind are responsible for their own ruin—their conditions of desertion having been self-generated from their own inability or unwillingness to make good on opportunities when others could (Biehl 2005). Rather than treating citizens as a single, unified, and collective body of subjects deserving collective security, citizenship entitlements are calculated under a logic of "variegated citizenship" (Ong 1999), whereby divided populations can be subjected to "different technologies of disciplining, regulation, and pastoral care, and in the process [are each assigned] different social fates" (Ong 2006, 7).

It is the emergence of these parallel architectures of care and abandonment that scholars note is among neoliberalism's greatest spectacles (Harvey 2006; Klein 2007). That it can maintain a sustained cultivation of zones of hyper-investment parallel with, and even at times adjacent to, zones of hyperneglect is indeed a peculiar feat because it begs the question of how it is that such stark landscapes of divided living—of hypervitalization and ghettoization—can evade collective alarm and indignation and come to pass as ordinary, commonsense geographies of the everyday. The logics of flexible citizenship suggest that it is the competitive actor's ability to have earned a place among those granted access to accelerated, intensified living—with its privileges of market investiture and flexible, spatial mobility—that sustain such spaces. What, then, to make of their extreme other?

For those still enabled with choices worth spending, at least, such spaces of abjection can be used to issue one valuable "pedagogical" lesson by governing power to individual subjects (Biehl 2005)—that of "using" and "spending" such options well. Such spaces of disconnection and exclusion, in other words, cancel the other's membership to "[make] it possible for one to belong . . . to a new population and subjective economy" (Biehl 2005, 65). It is, in other words, the vital fate of the excluded on which the explicit choice "for life" under neoliberalism extended. And thus through the public recognition and evidence of this logic does life, however selectively and narrowly construed, become "achieved through death" (Biehl 2005, 66).

Josefa Nolte's lament that "[ceramics are] an economic resource in a place where there aren't many resources" bespeaks precisely this because it acknowledges that however imperfect the conditions of its operation and however much it places at risk collective life in Chulucanas, opting to join the export market at least staves off for participants the fate of the unresourced and the unresourceful—that is, a future of not just uncertainty but also of irrelevance, a future of no future. And it does so by not only promising to prevent one from being designated as irrelevant and forgettable but also in offering the opportunity to achieve, to compete for distinction among fellow competitors, and to even perhaps be recognized as among their most successful, resourceful, and relevant members. Research has attended more carefully to the emergence and effects of these new dynamics of productivity and abandonment in settings where urban settings and zones of hypermodernity concentrate privilege, value, and worth. But this case should demonstrate how rural zones are increasingly the settings of such investments and impacts as well. And although the cosmopolitan urban settings of global cities that concentrate worth make evident what is at stake in successfully earning a place among the most globally productive, it is rural settings that arguably make more starkly evident what is at stake in "earning" a place among the unworthy—or worse, in choosing never to have competed at all.

Back in Lima, MINCETUR's Luis Calderón is describing for me how changes in techniques of production were part of the objective of the state's interventions in Chulucanas. We are sitting together in his sixteenth-floor office in San Isidro, Lima's financial district, looking over the knotted matrix of highways, traffic, and towers that spans before us. From his windows the city's impressions all weave together like a luminous web. Calderón, who directs the government's project in Chulucanas, recalls how directors at INDECOPI began planning a strategy for ceramics exports years ago, when it was decided that "what they had to establish were rules [that outlined] technical issues, [specifying] definitions and prohibitions, so that the product that they would develop would meet the demands of the market—not only in representing a recovery of techniques, but that would also innovate, that could earn them a place in the market, that would be sellable. . . . There are markets that are very demanding. One can use all the ancestral techniques, the materials, but still the market will see it as something crude, or rustic. . . . So it's necessary to look for an answer to this . . . to optimize plants . . . [and] to improve collections with technical assistance."

But in fact, there is little way to separate the rural artisan from the craft of production. And if the government's reform efforts hoped to modernize rural production, it would have to do so by modernizing artisans and cultivating enterprising "individuals." The operative phrases being deployed through initiatives such as these are the "capacitation" (or "training" and "reskilling") of individuals so that they are prepared to take responsibility for themselves and their newfound economic freedom. The key work of government here as in other contemporary information society and digital governance plans becomes not service to populations in general but service to select groups of individuals who can be retrained precisely so they can become newly responsible for what had been the responsibilities of government. That is, government's job becomes that of displacing and distributing responsibilities and seeing to it that "capacitated" individuals come to see themselves as responsible for project operability so that if information society projects fail, the conviction that individuals must then better realize their optimal potential becomes only further bolstered.

Indeed, faith in the Messianic global market in Peru today has proven hard to shake, precisely because it has generated new forms of productivity and productive subjects. Indeed, it has a new urgency beyond just Chulucanas today. In addition to requiring two new IP titles for exportable crafts per year, Peru also consulted with the Bolivian government to optimize its handicraft production. The promotion of Peru's national plan exportadora for agricultural exports and the passage of a free trade agreement with the United States were celebrated triumphs of President Alan Garcia's 2006–2011 administration. The Garcia government even contracted Peru's most prominent neoliberal philosopher, Hernando de Soto, in 2006 to see to it that its US Free Trade Agreement was passed.

If neoliberal logics have managed to engineer new geographies of wealth and productivity, it has not been without a cost, even for its most productive and active participants. The unsettlement expressed by actors such as Nolte over the lack of "social conscience" that's notable among the rural artisans who do secure export contracts today bespeaks precisely this. It likewise echoes a concern other critical scholars of neoliberalism have expressed before: namely, in the midst of these dynamic and massive transformations in "evaluating" and "valuing" populations, in precisely what exists community? Or as Jean Comaroff and John Comaroff have posed before: if among the "animating forces" of neoliberal capital are the impulses to "equate freedom with choice, especially to consume, to fashion the self, to conjure with identities; to give free reign to the 'forces' of hyperrationalization; to *parse* human beings into free floating labor units, commodities, clients, stakeholders, strangers, their subjectivity distilled into ever more objectified ensembles of interests, appetites, desires, purchasing power . . . [in] what consists the social? Society? Moral community?" (Comaroff and Comaroff 2001, 44). In the race to capture worth, recognition, and contemporary relevance and to secure a future of individual possibility, relevance, and consequence, in what does a future of "common relevance" and "shared consequence" get accommodated?

The state's acquisition of the DO demonstrates its recognition of Chulucanas as an exceptional site of cultural productivity, worthy of new and accelerated investments, and distinguishable among other rural zones that fall within national borders. But it also demonstrates the desire to reengineer Chulucanas into something more: not just as site of cultural productivity—where cultural knowledge is widely shared among artisans—but into a site of cultural innovation and authorship where design and techniques are instead carefully guarded. And a key part of ensuring the operability of Chulucanas's new IP title is the work of fostering new relations (or what we shall see become narrated as relations of promiscuity) that maximize individual potential and enterprise. These new forms of associativity, however, are coming into pointed conflict with those that preexisted it and were maintained by communities. If the case of Chulucanas demonstrates anything, then, it is that communities may have to be as concerned with successes of the state's new development initiatives as they are with its failures.

A display stand constructed by CITE at the Peruvian Gift Show features a selection of ceramics from Chulucanas surrounding a glowing tower of lights. The stand is just one among several displaying CITE designs to showcase Chulucanas crafts. The Gift Show, an annual event coordinated by the Ministry of Tourism and Trade at Lima's National Museum, aims to couple vendors of regional crafts with international exporters who arrive to the event from across South America, the United States, and Europe.

2   Native Stagings: Pirate Acts and the Complex of Authenticity

It is not even noontime yet and the television camera crews and photographers have already mounted their set-ups and stations in Chulucanas's crowded town plaza. They've arrived from the provincial capital city, Piura, and the even more distant national capital, Lima, to document the state-sponsored festival celebrating Chulucanas's new DO. The celebrations are more specifically timed around the recent establishment of the advisory board for the DO, called the *Consejo Regulador,* that will locally administer the right to use the IP title. It's this administrative body that will protect the exclusivity of the IP title, determining which local artisans meet the technical requirements necessary to adopt its use for their ceramics and which do not. And although much of the work of the Consejo is bureaucratic, the four-day festival in its honor provides an exhilarating visual banquet of regional culture that camera crews have traveled the distance to capture.

At the head of the plaza, a large stage has been erected to showcase performances of dances and music originating from Peru's northern coasts. Barefoot couples spin a jubilant tondero dance, whipping their ankles around in snakelike formations before a punctuated stomp drops to the ground. To one side of the plaza, diners sample the Piuran version of ceviche—featuring large chunks of regional fish bathed in the juice of locally harvested limes. And organizers weave between the crowds, handing out chilled samples of clarito, a sweet, northern coastal beverage (and cousin to chicha) drawn from fermented corn and traditionally served in a hollow, hardened gourd cask.

It is, of course, the ceramicists of Chulucanas themselves who are most prominently featured in the festival's events. MINCETUR and state officials eagerly shuttle between the stands of local artisans that frame the plaza. They stop to study the pieces displayed at each and greet their makers, flashing wide grins and extending ready handshakes to the craftspeople as a trail of photographers and onlookers follows behind. In the center of the plaza, a television crew from Lima leans their camera lenses in to tape artisans as they demonstrate the hand-molding and paddling techniques that are defining features of Chulucanas ceramics. Seated on the ground, one artisan carefully presses a mound of clay against the base of his bare foot, securing the damp material with his foot's curve while patiently paddling and turning it, a smooth stone cupped in one hand, until an evenly rounded bowl takes form. To another side of the plaza, a dozen or so artisans are bent over a table, each working furiously to sculpt blocks of clay into original designs during one of several live competitions in sculpting held during the festival. As town residents peer over each other's shoulders to catch a glimpse of the artists at work, a MINCETUR official stands to the side, waiting to triumphantly pronounce a victor.

Generating national visibility and public enthusiasm around the state's rural IP project and the native traditions it will be applied to, the event is by almost all measures a resounding show of success for INDECOPI and MINCETUR, the two government agencies responsible for developing Chulucanas's DO. These various stagings of indigeneity and resurrections of native traditions, in all their extravagant display, mean to draw attention to the important cultural preservation work that the government ministries and the Consejo pledge to undertake locally around the town's local traditions. Incorporated as central features of Chulucanas's new IP title, these production practices are part of the historical thread that links Ceramicas Chulucanas in their modern form to the ceramics-making practices of the pre-Columbian civilizations from which they originated thousands of years ago. By making such techniques an essential part of what should be considered in deciding which artisans will be granted the rights to adopt the DO for their products and which will not, the Consejo also aims to preserve Chulucanas's ancestral traditions. And by staging such traditions, prominently, the state can pronounce and affirm its newly intensified dedication to cultural diversity and pluralism.

This, at least, is how Peru's government agencies intend the public to understand their work and the Consejo's in the events celebrating the IP-based initiative. Having spent the past several months interviewing and traveling with officials from MINCETUR and INDECOPI responsible for developing Chulucanas's DO, it's an explanatory discourse and rationale that I've become accustomed to hearing, which is what makes the news I hear from local artisans shortly after the festival seem all the more incredible.

In January, just a few months before the festival itself was scheduled to begin, an international scandal involving a major US home décor chain and hundreds of thousands of dollars' worth of export goods was traced back to Chulucanas. The US company discovered that a shipment of traditional pottery received from Chulucanas had been faked. Thousands of ceramics pieces that were supposed to have been kiln smoked to acquire their darkened color had instead been painted black. The tags that would have been attached to the ceramics to vouch for their authenticity—and that described the techniques traditionally applied for production, including smoking with mango leaves and hand and paddle shaping—were left aside. Black paint, however, was visibly peeling off ceramic surfaces and sticking to the palms of handlers. At the height of the US holiday shopping season, while shop floors were frantically stocking shelves with merchandise for eager consumers, the retail chain realized it would have to discard the entire shipment and seek answers from the Lima-based exporter who had contracted the order.

None of this, however, is publicly mentioned during the festivities that unfold in Chulucanas just a few months later. It is not mentioned to me by any of the government stewards whom I had begun to interview months previously on the work of developing the town's DO, even in the course of traveling together between Lima and the small town in the country's far north. Neither was it raised to me by any of the artisans who had been working most closely with the state in the IP-based initiative and with whom I had had routine contact. It's not until well after the festival, in fact, that I first heard of the hoax itself and learned of the enormity of the stakes involved—not just for the actors directly implicated but for the entire population of Chulucanas artisans. A scandal of this unparalleled size placed in question the town's reputation as a reliable producer of global exports and threatened the region's international branding. It put at risk not just this particular US contract but future export contracts as well. And fewer contracts meant in the end fewer jobs for residents. In the midst of the state-directed celebrations around Chulucanas's ancestral traditions in ceramics making, the village's largest sales hoax in traditional exports had somehow been committed. And even as the details surrounding the magnitude of the scandal and its responsible parties were still unfolding, state workers were channeling efforts—not into answering the question of how it could have happened at all or how to prevent its repetition but into fanning the festivities celebrating the town and its history of honored tradition.

As puzzling as this might sound, more confounding still is the notion that the very government stewards who claimed to work for preserving Chulucanas local arts would at once explicitly refrain from issuing a condemnation or even public recognition of the incident. Such a public condemnation or even simple recognition—whether of violating cultural tradition, transgressing community norms, or breaching business con- tracts—in fact, would never come. It was as if the incident of global cultural fraud had appalled everyone except those public officials most actively involved in the contemporary promotion and protection of Chulucanas ceramics. It was as if the incident was not seen as worthy of condemnation or mention at all or had been seen on some not-so-distant register as unex- ceptional. Or perhaps, it had simply been anticipated all along.

Staging the Native

A month later in Lima, Chulucanas ceramics tradition is again being promi- nently staged. Just a few weeks after hosting the local festival in Chuluca- nas, MINCETUR opened a special exhibition in The Joaquin Lopez Antay Gallery of Traditional Peruvian Art in Lima's downtown historic center cel- ebrating Chulucanas ceramics. The exhibit draws together a selection of early and contemporary pieces from some of the most renowned workshops and artisan families in Chulucanas. There is a hand-molded crucifix from the workshop of Max Inga, one of the first ceramicists of Chulucanas to achieve global recognition for his craft, who passed away just before Chu- lucanas's large export orders began production. There is a large example of a *chicera* sculpture, representing a local female chicha brewer surrounded by a ring of ceramic urns used to store the traditional beverage that Gera- simo Sosa's workshop became famous for decades ago. And representing the more contemporary versions of Chulucanas ceramics are pieces from the workshops of Jose Sosa and Santodio Paz, including vases produced and shipped for some of the most recent export orders.

At the opening reception for the exhibit, I find myself inside the crowded gallery, peering over the shoulders of a fleet of news camera crews and pho- tographers, trying to find a clear view of the directors of several government ministries who are at the center of the media-documented commotion. The exhibit is one of the main highlights for MINCETUR's Craft Week, and the government officials—including the very photogenic Mercedes Araoz Fer- nandez, the minister of trade and tourism—kick off the event by praising Chulucanas ceramics as uniquely representative of Peru's treasury of indig- enous crafts. Indeed, as the first traditional craft to have acquired a DO and the first in the country to have been honored as a national product, Ceram- icas Chulucanas had managed to uniquely distinguish itself in a nation whose traditional arts made it, as one exhibit plaque proudly declared, one of the "richest on the planet." Another gallery plaque elaborated on the production techniques that date back to 500 B.C. maintained by Chuluca-nas artisans. Paddle molding, clay smoking, and positive and negative col-oration have been practiced since the era of the pre-Columbian Vicús and Tallan peoples. These practices endow Chulucanas ceramics with "a sin-gular beauty" that has made it a "trademark of our country in the world."

With some $40 million USD in annual exports, the popularity of Chu-lucanas ceramics among international consumers seemed to have earned it the reference of "global trademark." This is in addition to the fact that Peru's global craft exports had already been identified by the state in its 2003–2013 National Strategic Export Plan as a key sector of economic growth. Having increased some 62 percent in sales between 2009 and 2010 (Andina 2010), craftwork served as an exemplary model of the very "export culture" that the state specified it sought to cultivate among national popu-lations. Such national projections echoed the similar global trends that the UN Conference on Trade and Development (UNCTD) had recently tracked. In its 2008 *Creative Economy Report*, arts and crafts were framed as "the most important" creative industry for developing countries, making an area in which developing nations had a leading position in the global market and for which exports nearly doubled between 1996 and 2005, from $7.7 billion USD to $13.8 billion USD (UNCTAD 2008). And explaining Peru's valoriza-tion of the concept of "cultura exportadora" ("export culture") for economic growth in the nation's own "master plan in export culture," MINCETUR's Alfred Ferrero wrote, "In each step of the export chain is always work con-nected to an Export Culture [of] efficiency, competitiveness and productiv-ity. In MINCETUR, we have tried to go even further and develop guidelines and public policies for export culture. . . . A culture of this nature should lead us to see export as an activity that marks the economic fate [of] our country. This should be based on merging values, modes of life, customs, knowledge, and level of development favorable to *competitiveness*" (Gobi-erno de Peru 2006, 6 [emphasis added]).

That a provincial town in the rural northern stretches of the country could be an exemplar of such competitiveness, responsible for millions of dollars of global exports, and capable of winning the hearts—and invest-ments—of consumers worldwide was an achievement that the Lima-based attendees of the Craft Week event considered visibly praiseworthy. That humble artisans from one of the country's innumerable rural villages could compete with the likes of other newly industrializing export giants such as India and China and could generate growing international demand for an artifact that, furthermore, was inextricably bound to the ancestral, indigenous origins of the nation would have aptly stirred the pride of any compatriot—even if he or she personally had never visited Chulucanas before, as was the case with most of the event's audience. It's in large part this consciousness of the nation's history of indigenous civilizations and a general awareness that Chulucanas artisans represent only a small portion of the countless other artisan populations around the country's rural territories that fuels a palpable hope around Chulucanas ceramicists and the new developments around the town's DO. If Chulucanas artisans can succeed in having their native arts and traditional crafts valued by global consumers, so, too, could countless other rural populations scattered throughout the country. To the urban, capital city residents attending the exhibit, national economic development could seem as close as simply making such diverse regional resources of indigenous culture globally visible and accessible.

As much as the MINCETUR exhibit was a celebration of Chulucanas artisans, it was also a celebration of the international market channeling these new valuations—not only for Chulucanas local products of the various other cultural traditions still actively practiced in Peru but for the modern, multicultural Peruvian nation as a whole. It's this that is particularly highlighted in the exhibit's commemorating inscriptions of Chulucanas traditional ceramics as a newly market-recognized and contemporarily valued "trademark of *our country* in the world." Captured in such inscriptions is an image of today's global market as having afforded Peru's regional arts a new kind of global mobility, circulation, and visibility and as having recovered the nation's native crafts and allowing them to acquire a new and present desirability. It is, after all, this market that has provided new esteem for Peru's native arts, enabling them to accumulate new scales of financial investment that thousands of years of anchored, domestic existence (and largely marginalized at that) had never afforded it. And it was this international market that—presumably—had at long last recognized the intrinsic value of Peru's native arts, lifting them from public obscurity and granting them new scales of vibrant, international visibility.

The contemporary market, through such state framings, is thusly endowed with a salvational potential that scholarly critics of neoliberalism have noted before. Promising to provide traditional artisans with the financial rewards that they had unfairly been denied before, it is this market that not only promises to provide new tribute to artisans that can make up for past historical deprivations but that also pledges to secure the reproduction of their traditions for future generations to come. In so doing, global markets demonstrate a stunning capacity to lift rural villages from a fate of eternal condemnation to poverty and to miraculously recover them instead into historical relevance.

But although neoliberal states have been enthusiastic promoters of the global market as the salvation route for the rural poor, they've been less eager to acknowledge how they too can benefit from such rescue acts and assuage their troubled history of governance in the postcolonial, pluricultural, contemporary culture. Global markets are able, supposedly, to newly recognize and reward value in the cultural expressions of the very traditional and native populations who had historically experienced government neglect, and grant states the chance to unshackle themselves as well from a legacy of colonially rooted, Western-centrism and the prejudices they spawned. If colonial logics of racial exclusion still visibly haunted postcolonial states—producing devalued indigenous cultures, marginalized rural populations, and, ultimately, a fragmented, uneven process of nation building in Latin America (Dussel 2000, 2002; Escobar 2007; Quijano 2000, 2007), then coupling with diversity-incorporating global markets provided a shot at redemption from such a past. It allowed an opportunity for states to turn away (at least temporarily) from the familiar logics of exclusions cultivated under earlier forms of Western universalisms and provided a chance to reestablish ties to local, provincial zones.

Moreover, they offered a chance for states to be resurrected as governments that newly acknowledged the error in previously undervaluing traditional culture and communities and could instead embrace modern values of liberal, democratic citizenship, multiculturalism, and pluralism. Significantly, however, if such contemporary resurrections allowed states to be redeemed with a new sense of justice and inclusion, they did so while also incorporating an economic calculus of rationality efficiency. Reborn as an improved, market-rationalized twin, the neoliberal state could bestow its benefits selectively, rewarding cultural diversity and value when it could be duly acknowledged—and withholding compensation from those deemed aptly undeserving.

But in all the celebratory euphoria of the museum event in Lima and elated recognition of the cultural recovery work newly performed under the contemporary market's meritocratic valuations, there were a few absences. The striking changes in ceramics production that unfolded with the town's explosion of export orders and that have been part of the modern history of ceramics making in Chulucanas were omitted. There's no mention of the new technologies and production techniques that global exporters introduced into Chulucanas in just the last decade to massify and serialize production for large exports. And there's no distinction made for audiences between the works exhibited by Chulucanas's earliest recognized ceramicists and those representing contemporary production for the mass market.

Although the oldest ceramics displayed represented work by artisans whose reputations were built through individually crafted pieces using traditional techniques, the contemporary selections included were made by artisans who no longer applied traditional production methods and whose reputations were founded on the large-scale export of serially produced pieces.

The omissions are all the more curious given how the transformations in Chulucanas ceramics production are arguably what led to its being awarded an IP title with the DO. It's these transformations in production that, according to many actors who had witnessed the changes in Chulucanas unfold in the past decades, gave the once localized ceramics a new global life, making it internationally visible and recognizable and converting it from a product sold in local markets and with limited urban distribution into an world-circulating good whose global distribution had increased exponentially. If until the 1970s, Chulucanas ceramics were relatively rare objects, existing only as unique, handmade pieces that sold for around $150 USD in a few boutiques in Lima, Europe, and the United States, by the late 1990s, it was no longer the same artifact. Readily available in Lima's budget-friendly and high-end markets and in US chains as distinct as Neiman Marcus and Ten Thousand Villages, Ceramicas Chulucanas in just a few decades had gone from public obscurity to becoming one of Peru's most recognizable folk arts.

Among those who share an intimate awareness of how deeply these transformations in ceramics production have affected local life in Chulucanas is Luiggi Castillo. As the director of the CITE Ceramicas, a MINCETUR-managed agency based in Chulucanas, he has dedicated his work to operating an institution with a more than twenty-year-long history of working with local ceramicists. In recent years, CITE's primary activity has consisted of encouraging artisans to incorporate new technologies into their workshops as a means of improving product "quality" and enhancing Chulucanas's competitiveness in the global market. He makes a distinction between the "improved" quality of exportable ceramics, for instance, and those that are sold in local markets, crediting the success of Chulucanas's global exports to improvements in product quality that CITE's worked to promote: "Year after year, the demand for ceramics from Chulucanas is greater, so more people have gotten involved in ceramics production. And what we're seeing is that this Chulucanas ceramics can achieve the requisite quality standards so that the product doesn't weaken in the market. . . . There are different qualities [between ceramics from workshops]. There are those who work very well and those that just work for a small market [*mercado bajo*] . . . where the quality is very low . . . [and] they don't reach good quality standards—optimal standards. Whereas it's the export ceramics that demands the highest quality."

Appointed to serve as the head of the Consejo Regulador for Chulucanas's DO, Castillo explains that it's the promotion of these technical standards that will bridge the reform work of CITE Ceramicas with that of the Consejo: "This is why we've proposed the Denomination of Origin, why it's important that it's applied—to establish what the technical norms for Chulucanas's ceramics should be and establish a manual for best practices. . . . The Consejo Regulador will allow us to administer the brand and authority of the denomination of origin to soliciting workshops." He stresses for me that it's this "quality control" work of the Consejo that endows Chulucanas's new IP title its true power, encouraging local artisans to adopt technical reforms as a requisite to being able to use it. "To be able to claim the Denomination of Origin, these workshops should be able to work under the technical norms that the Consejo Regulador demands, under the authorized conditions that will produce Chulucanas Ceramics."

Castillo's explanations, similar to the state's cultural stagings, testify to the preservation work that Chulucanas's IP-based project will allegedly undertake. Crucially as well, they are intended to testify to the salvational power of contemporary global markets and their capacity to reward the value of native indigenous traditions that had been previously unappreciated. But Castillo's narrations of the transformations traditional workshops are encouraged to adopt underscores a key tension in advocacy around the DO and its claims to native cultural preservation. Although the state's IP project and the promise of national and regional economic development that are advanced with it have been publicly celebrated for their capacity to protect and promote native cultural traditions, its accommodation of large-scale, nontraditional production methods enables it to advance other interests at once. In its local implementation, it has highlighted the need for transformed techniques and proof of "quality" craft making before workshops can be entitled to use the IP title. In practice, local artisans aren't granted the right to the DO because of their history of local production or maintaining traditional techniques but on the earned ability to demonstrate the "technical norms" required by the Consejo. In the name of optimizing Chulucanas ceramics for large-scale production and global ability to circulate, Chulucanas's IP title legally authorizes (and arguably obligates) the remaking of native crafts themselves. And although it is the protective work around native tradition that is most prominently publicized in advocacy for the DO and its new markets, it is the unspoken work of displacing native tradition that has left its local mark in Chulucanas.

But if the narrative of the salvational global market masks a local reality, it does so only partially. Against the explicitly public narratives and stagings that celebrate a new salvational market and the redeemed, multicultural state that now rewards traditional culture are the quiet but critical counternarratives of local artisans. Such narrations challenge claims of cultural preservation made under the state's IP project by critiquing contemporary ceramics production in Chulucanas and stressing the degradation of actual conditions supporting traditional expression. Importantly, these counternarratives demonstrate how the social ties that had helped to sustain local, indigenous tradition have been compromised in the midst of promoting export networks and even weaken "technical standards."

Such narrations give voice to the incompleteness of the global market's persuasive power as well as that of the neoliberal policies advancing it. Even as neoliberal promoters insist on the liberating power of global markets and their ability to revalue local culture, artisans' accounts of social disintegration stress its destructive capacity instead. Although pro-export actors appeal to a religious-like faith in global capital's redemptive power and its capacity to right past wrongs to recruit participants, artisans decry a betrayal of community ethics, custom, and the traditions surrounding native cultural production and expression to prevent such partnerings. Through the critical accounts circulated by artisans, the state's stagings of indigeneity appear not as authentic expressions of a contemporary will to preserve tradition but are exposed instead as strategically engineered essentializations whose purpose leans more toward logics of exclusion of undeserving partners than broad cultural inclusion.

Against the salvation stories surrounding the global market then emerge another set of hypermoralizing tales. These accounts renarrate the global economy as cultivating not merely relations of networked culture but also flexible relations in which diverse parties—including ones with oppositional interests—are brought into strategic allegiance. It's this web of relations that invites culture and tradition to be rerendered into new versions of themselves that can be globally produced and circulated with frictionless efficiency.

But even for the willing partners of the global market, who readily provide indigenous goods that attest to tradition's ability to be authentically captured and valued through global export, such smooth efficiency, similar to the salvational tales of global capital, prove impossible to consistently stand by. By the individual accounts that artisans locally circulate, offenses such as that of the painted pottery shipment have not only amplified the scale of potentials for cultural fraud but have also increasingly found a place in Chulucanas's production landscape as commonplace elements. What might have once occurred as rare, isolated incidents of production error whose distribution was limited to a local sale are narrated as now assuming global proportions. Such tales on the growing incidents of transgression evidence a recognition of the unfulfilled promise of salvational capital. They highlight local producers' sustained will to give voice to the unspoken local impact of export-oriented developments. The persistence of such disclosures further demonstrate artisans' condemnation of practices of cultural fraud whose expansion and incorporation into the ordinary has quietly, and almost incredibly, escaped public detection. They remind us, too, that not so far beneath the smooth, public surface of the promises of redemption and salvation by global capital and the multicultural state lie deep fissures where the narratives of the unrecognized and unvalued pocket, discreetly accruing until the ruptures can no longer be contained.

Exclusive Indigeneity

Cultural theorists have observed that the origins of indigeneity as a modern, strategic performance lie in the late twentieth century when *culture* became increasingly turned to as an "expedient" to solve problems of social conflict or development that once were the province of national economic and social policies (Yudice 2003). Whereas the question of culture and problems of socioeconomic integration were once state responsibilities that called for the expenditure of government resources to address the shifting valuations around diversity—and indigenous identity, especially—reframed culture into a contemporary resource that could be deployed to capture new market values or leverage nation building. Scholars have noted how the United Nations' issuing of the Universal Declaration of Human Rights (1948) contributed to the positioning of diversity and pluralism as key international interests that should be universally adopted (Goodale and Merry 2007; Keck and Sikkink 1998). Such values as well as an international imperative to promote the world's cultural diversity and protect endangered minority populations were concretely issued through a variety of conventions in the post–World War II era, including the UN's International Covenants on Economic, Social and Cultural Rights (1966) and on Civil and Political Rights (1966), and the International Labor Organization's Convention No. 169 (1989). The global interest in conservation expanded even further to include not merely human diversity but also all forms of biological and genetic diversity when the Convention on Biological Diversity was issued in 1992. Such political documents provided powerful discursive tools for shaming governments who had subjected indigenous populations to centuries of political discrimination and marginalization (Jackson 2007; Niezen 2003; Speed 2007).

Scholars of contemporary social movements have noted, too, how indigenous movements have leveraged such discursive tools in a range of political struggles in the latter half of the twentieth century. Fighting to address problems that have persisted since the colonial era, including the loss of native territories, the breaching of legal treaties, and the imposition of socially and culturally destructive assimilation policies (Bebbington, Connarty, Coxshall, O'Shaugnessy, and Williams 2007; Niezen 2003), collective actors have strategically leveraged "universal" values of Western liberal pluralism to issue critiques. Yet by circumscribing states' special obligation to protect traditional peoples and the survival of their "distinct ways of life" that separated them "from other sections of the national community" (ILO 1989), these international declarations placed a burden of proof on populations themselves—obligating them to prove their authenticity and "distinction" in order to make claims of political rights or entitlement (Graham 2002; Jackson 2007; Montejo 2002; Povinelli 2002; Turner 2002; Warren 1998; Warren and Jackson 2002). Despite intending to empower traditional peoples, such doctrines instead cultivated conditions in which the performance and assertion of cultural difference—however self-essentializing and romanticizing—were demanded as requisites to address states. The successful enactments of an indigenous self as a surviving form of ancient, primordial identity (even when not recognizable to community members) emerges, then, as a condition for receiving institutional recognition and benefits. Elizabeth Povinelli (2002) writes compellingly of the impossible demand made of such performances: "In these uneven cultural fields, how do jurists and other nonindigenous citizens discern a real indigenous subject from a 'more or less' diluted subject? Is it sufficient for indigenous persons to assert that they know customary beliefs or must they demonstrate some internal dispositional allegiance to that belief? How does one calibrate an internal disposition" (4)?

More recently, scholars have pointed to the new valuations that indigenous resources and traditional cultures have been granted through contemporary market forces (Bellman, Dutfield, and Melendez-Ortiz 2004; Brown 2003; Comaroff and Comaroff 2009; Coombe 1998a; DeHart 2010; Dutfield 2000; Hayden 2003; Posey and Dutfield 1996; Riley 2004; Shiva 1997; Suthersanen, Dutfield, and Chow 2007; Yudice 2003). As guardians of knowledge and culture that could feed growing consumer demands worldwide for traditional products, native populations have been urged by various private actors—from pharmaceutical companies seeking to derive new drug treatments from traditional medicines (Hayden 2003) to marketing companies seeking to use native symbols for product advertising (Coombe 1998)—to partner with commercial actors to commodify traditions. Such "cooperative" relations promise financial rewards from global sales to isolated native groups who can be identified as the "proper" author or owner of tradition. These market-based compensations obligate native groups to acquiesce to the authority of modern state's property law and the establishment of exclusive rights to authentic owners and authors, even when the traditional form in question had been collectively held or widely adopted among many native peoples (Shiva 1997).

Although asserting themselves as working in the political and economic interest of indigenous peoples, then, the new valuations of indigeneity have operated with a fetishization of "authenticity" that can serve to disenfranchise native peoples. By establishing an objective, verifiable litmus test for "authenticity" on which to base political recognition, the withholding of rights and benefits to native populations can assume a coolly rational veneer. Framed as a naturally scare resource, authenticity can be seen as a means to exclude those parties who fraudulently claim political rights and economic benefits. Reifying antiquated notions of native identity and culture, as well, and ensuring authenticity has encouraged the adoption of what social movement scholars have called practices of strategic essentialization, in which ironic and self-conscious performances of "traditional" identities are deployed to make rights claims to states. But the same logic has placed native peoples' claims to indigeneity subject to intensified scrutiny, producing a rationale through which rights, which often had been historically withheld from native peoples, may find a new mode of being "legitimately" denied.

Still, it is not native peoples alone who find themselves confronted with the dilemma of performing authenticity. Faced with the possibility of having the long-marginalized native peoples they've worked with as colleagues and partners disenfranchised yet again, academic and activist spokespeople for native groups have been obligated to fraudulently represent themselves, too, maneuvering around a language of "authentic" indigeneity and culture in parallel practices of strategic essentialization (Povinelli 2002). Much as global researchers seeking to attribute local origins to traditional medicinal plants to be used in drug design must work effectively to produce such sites and "authors" of origin (Hayden 2003). More troubling, perhaps, is how authenticity's calculus for rights granting actually can manage to portion out benefits. Adhering to a logic that a proper native beneficiary does exist, authenticity's fetish generates conditions whereby native groups must compete for exclusive recognition as the "true" and rightful beneficiary, producing conditions in which collective benefits and potential solidarity are undermined.

The state's own interest in producing native stagings, and indeed in benefiting from its own practice of strategic essentializations of indigeneity, then prove central to its information society initiatives. By justifying a logic of conditionality for the provision of rights based on citizens' ability to productively use and deploy resources of the information economy, states convert rights themselves into contingent and exclusive provisions. Relieved of the obligation to generalize resources, states can instead conserve benefits for those deemed most deserving. Such a logic shifts obligations from the state to benefit-seeking groups, who must now prove why investments should be rightfully channeled into them and to the exclusion of other parties. But it is more than a question of the efficient conservation and distribution of resources in the native stagings states engineer. The Peruvian state's stagings—integrated into UN-sponsored global conferences and televised national art exhibitions—demonstrate how they are produced in the interest of maximizing public exposure and international visibility.

And little wonder why. At stake, too, is the performance of states as contemporary political actors that can effectively defend "universal" values of multiculturalism and inclusion in the information society. For Peru, as for many postcolonial states in the Americas, the demonstration of a commitment to such liberal values of pluralistic, social inclusion—particularly of formerly excluded rural and indigenous populations—was central to nineteenth-century nation-building projects (Joseph and Nugent 1994; Nugent 1997). Contrasting the liberal, egalitarian values of newly independent states against the oppressive rule of colonial powers was a means for new governments to distinguish themselves from former rulers, gain legitimacy among a fragmented populace, and ultimately to claim official statehood.

Yet today, the mounting incidences and intensifying grievances of traditional populations across the globe demonstrate that pluralism's practice remains problematic. In the face of such conflicts, the ability to provide public evidence of the embrace of multiculturalism as an official national value—however partial such an "embrace" may be—becomes all the more imperative for modern states. Harnessing the emotive power embedded in selectively engineered acts of pluralism, then, is an essential element to such stagings. Through such work, states may—cunningly—provide the national public with objective proof and experiential evidence of their actualized good will toward diverse actors. In so doing, they at once affirm their ability to act in the name of the "authentically" modern, multicultural nation as much as they resecure the public authorization to act on its behalf. Or as Elizabeth Povinelli (2002) writes of the Australian state's own multiculturalist performances: "As the nation stretches out its hand to ancient Aboriginal law, indigenous subjects are called upon to perform an indigenous difference in exchange for the good feelings of the nation and the reparative legislation of the state. But this call does not simply produce good theater; rather, it inspires impossible desires *to be* this impossible object and transport its ancient, prenational meanings and practices to the present in whatever language and moral framework prevails at the time of enunciation" (6).

Although scrutinizing the authenticity of indigenous performances as a means of portioning benefits continues unabated, subjecting the authenticity of states' performances of pluralistic inclusion to an equivalent scrutiny has yet to begin. Such uneven attentiveness to authenticity of performance and how contemporary multiculturalist acts of modern governance have managed to escape the same kind of public scrutiny indigenous performances have received, indeed raise their own questions. In a truly pluralistic state, why would native groups resort to such performances as a means of gaining economic or political security? Why, unless the actual survival of such groups were at stake, would the strategic uses of an essentialized culture have even become necessary at all? And why, in the strangest twist of all, would traditional groups have to compete among themselves for authentic entitlement, as if authenticity itself were truly a scare resource?

Export Redemption

In the photo that appears of Maria Carmen de la Fuente in a splash article in the national newspaper *El Comercio,* she's seated nobly, her hands folded in her lap, in a room whose walls are lined with several dozen pieces of Chulucanas ceramics. It is a fitting image. For more than twenty years, the fifty-nine-year-old founder of the export company ALLPA has directed an enterprise whose own history is inextricable with Chulucanas ceramics. ALLPA was the first export company to recognize the global market potential in Chulucanas's locally crafted ceramics, dedicating itself to promoting the local production transformations that were necessary before the traditional good could be turned into a global export. And it was ALLPA that first proved in the mid-1990s that Chulucanas indeed could realize such projected potentials after successfully contracting the town's first large order of twelve thousand ceramics for the US home décor chain Pier One.

It's this heroic thread in ALLPA's history with Chulucanas that's empha-sized in the headline for the interview with the exporter published in *El Comercio*'s business section. In bold print, just to the side of de la Fuente's winning smile, it pronounces: "She's about large-scale production, and she has alliances with thousands of artisans. At first, her project was criticized. Now she and her artisans are giving Chinese [producers] a run for their money." It sets an appropriate tone to an article that means to narrate the triumph of the free market to bring progress to Chulucanas's poor artisans, who, despite some initial resistance, were successfully transformed into productive subjects, now able to compete in the global market against even the formidably competitive Chinese.

Since ALLPA's success with Chulucanas exports, it has branched out into several other rural zones known for their traditional and artisanal crafts. But De la Fuente cites Chulucanas as the company's original—and larg-est—success story, the one on which their initial fortunes and growth were founded and the one that would provide the model for reproducing simi-lar market successes in the multiple communities where their work now extends. She doesn't mask her pride when she speaks of Chulucanas, of ALLPA's labor there to develop an export market, and in particular, of the "the miracle" that she says unfolded there following the company's invest-ments. Highlighting the salvational power of global markets as not just able to redeem Chulucanas's impoverished artisans and markets but also able to do so with a magical, unimaginable power, she tells me, "The miracle came about almost in the first or second year—this miracle that brought monstrous export orders that the artisans had never before in their lives dreamed of." Embedded in de la Fuente's framing of Chulucanas's export "miracle" is a significant evolutionary narrative. If global capital's salva-tional potential could "redeem" Chulucanas's producers, it was implicitly because of their own sustained failures and underperformance—offenses evidenced in their inability to have made good on the market potential that they had seemingly always possessed but that remained unrealized until the interventions of export companies such as ALLPA.

De la Fuente emphasizes the range of these investments in her inter-view with *El Comercio*, speaking of the product designers and engineers who ALLPA helped bring to Chulucanas over the years to identify existing production inefficiencies and introduce reforms for amplifying export pro-duction. One of the earliest and most significant production inefficiencies identified was artisans' use of the paddle and stone to give form to the raw clay. Such a technique rendered individually unique pieces that, although acceptable for individual sales, were considered product errors in a mass market that demanded numerous goods identical in size and shape. More significant was that traditional production techniques required that an artisan spend nearly an hour of graduated, careful paddling before a piece was given its finished form—a process that limited artisans' production potentials to no more than several dozen pieces a week.

To correct such inefficiencies, or what de la Fuente refers to as Chulucanas's "primitive" technological state, ALLPA proposed replacing the traditional paddle and stone shaping with the electric potter's wheel. Such a change would ensure not only more consistency in ceramic forms but would also vastly accelerate production potentials, allowing a single workshop to increase output from several dozen to several hundred pieces a week. It was a radical proposal, not only because few ceramicists in Chulucanas had ever seen, let alone used, a potter's wheel before but because it would also mean disregarding traditions shared among all of the area's living artisans and that bound them to thousands of years of past generations of local potters. ALLPA would be unable to begin putting their reform plans into action, in fact, before first importing workers into Chulucanas who were already skilled in using the electric potter's wheel for serialized production. Drawn in from cities all across Peru—from Lima's coastal capital to the central Andean mountain cities of Cuzco and Arequipa—this new group of knowledge workers was transplanted into the town for the precise purpose of teaching its skill to local producers in Chulucanas and enabling the creation of a new local class of laborers proficient in the techniques of massified production.

For all of de la Fuente's public assurances of the export market's possibilities, she recalls that artisans still reacted with pointed indignation, attacking ALLPA's export-oriented reforms as a "sin" against tradition. She remembers how she had to cite the Old Testament as proof that the potter's wheel was used since "biblical times" in attempts to convert artisans into reform believers. She recalls, "When we first arrived with the potter's wheel, they looked at it and thought it was a strange machine because it revolved. We had a lot of problems with the most traditional artisans because they thought it was a sin, a tool that was going to replace the paddle that had been used since pre-Incan times, for the potter's wheel. We told them that the wheel was a tool that was used since Biblical times. Because in the Bible it's cited that the wheel was used by potters."

De la Fuente remembers that artisans' resistance only grew more pitched after production for the first large export orders ceramics began in Chulucanas's first modernized workshop. The situation would become so tense, in fact, that the town would divide along pro-export and anti-export lines.

"At first there was a reaction that was a little ugly from the town. From the competitors that weren't involved in the 'miracle' because they saw that the orders were very large, that buyers were coming, that companies were coming but that it wasn't possible to work with everyone. So those that weren't benefited with orders developed a kind of *war against exporters*." But de la Fuente's seemingly contradictory account of anti-export warriors as representing "the town" in its entirety and as being composed of only those artisans "who didn't benefit" from export orders is interesting precisely for its counterintuitive accuracy. ALLPA's critics could at once encompass nearly all of Chulucanas and be isolated to only those who didn't benefit from the town's first export order only because of the limited size of those who did benefit. And de la Fuente's slippage in describing her opposition as the "entire town" and only a select group of nonbenefitting artisans demonstrates her awareness that profits themselves concentrated in the single workshop the company first selected to modernize.

She recalls that these artisans would persist in their condemnations of the ALLPA project, circulating a moralizing discourse that positioned the production reforms as an assault on collective life in Chulucanas—in its past and present form: "For a while, they maintained their attack . . . attacking us because we had brought the potters' wheel, because they said that this was a damned machine—and they said just that, a 'damned machine.' That it's a robot that's going to automate production, and that it would generate unemployment in Chulucanas. That thanks to the presence of that machine, there was going to be unemployment and that they had to maintain the traditional technique of the paddle because that's what all our ancestors did."

Despite such pronounced resistances from local producers, de la Fuente maintained her faith that they might still be converted. Explaining how a firm belief in the untapped, neglected export potential in the town resiliently anchored her commitment to her work there, she told me: "The thing is, Chulucanas is very backwards, but we saw that Chulucanas's ceramics had an immense potential, and that it was necessary to modernize it, modernize it in the technical part, and modernize it in design, in the order of production, and this was what we did in those first years, and it's what we still do." Indeed, the work that ALLPA initiated in Chulucanas when it first began to develop the town's export market years ago continues today in a significantly expanded version. The company continues to invest in modernizing artisans' workshops and organizing local training workshops run by design and engineering consultants from Lima, the United States, and Europe to introduce artisans to new technologies and techniques to boost production potential. The company had partnered with just one artisan workshop back then but today it partners with seven or eight workshops. And while its first major export order was of some twelve thousand pieces, by the new millennium, its orders climbed to five times as large. It's the export company's local reception, however, that most tellingly bespeaks for de la Fuente the scale of transformations in Chulucanas and the successful conversion of artisans. Once the object of community revile and condemnation, the company today receives countless inquiries from local artisans seeking to partner with them. She reports, proudly: "When we started, there was a division between the purists, those that spoke of not transforming the craft, and maintaining traditional techniques, and us, who proposed serial production, mass production. Many people criticized us. Now no."

Despite its years of profit-generating investments in Chulucanas, however, and the successful recruitment of multiple new artisans as partners, de la Fuente stresses that its work to optimize the town's traditional production methods is far from over. Several weeks after celebrating the recent contract of seventy thousand ceramics for Pier One, she has already turned her focus on how to fix the imperfections and inefficiencies she sees as still evident in today's workshops. She tells me: "There are always inefficiencies. There doesn't exist an artisan workshop that functions optimally because there are always inefficiencies . . . especially when you work with large volumes, because when you work with small volumes, there aren't big problems. . . . [But Chulucanas's artisans] have been left in a technological level that's very backwards, so . . . they aren't prepared for large orders where standards are more and more challenging. Now it's not like it was ten years ago. Now, for example, in this order from Pier One, they ask for a whole lot of requirements that they didn't ask for five years ago. The market is more demanding. So the artisans keep being people with a low level of instruction."

De la Fuente emphasizes the range of these investments in her interview with *El Comercio*. "What there is here is a lot of effort, and the dedication of a team of workers. We are 26 people that work 11 or 12 hours daily, because we invest a lot in fair participation, in designers, in consultants. Because it's not just that we have to improve the product, we have to also see how to attain a competitive price." But oddly, it's precisely all this labor to optimize and transform ceramics production and pricing that remains invisible on the showroom sales floor. There, exporters' investments are meant to be channeled into the showcased product for one final and all important staging. Framed for clients as a native good that resurrects the indigenous traditions of Peru, it must be able to mask the displacement of traditional production techniques that, indeed, exporters have labored so hard for.

And it must be able to perform its natural authenticity with the validation of export specialists who have dedicated themselves to the search for and market delivery of such native goods. Under such sales-oriented native stagings, native products are scripted as naturally occurring artifacts, as having been found rather than transformed, and as objects that were "discovered" rather than "remade" by exporters. Or as De la Fuente describes it, "On the showroom floor, our product almost sells itself. . . . We've put a tremendous effort into offering a product that's beautiful and well-priced, a product that's backed by a serious company. That's why I say that our product sells itself. In a fair, I describe to the client how it's made, and that we have 20 years in this business. . . . What you've got to do is *transmit authenticity*."

Authenticity's Complex

I met Miguel Pachas for the first time at his home in Chulucanas late one afternoon. It is the end of the workday when I arrive, and I see the last few of twenty-five or so people he currently employs leaving the workshop that operates in a space just to the side of the Pachas family house. The workshop is in the middle of production for a large order for ALLPA, which ceramicists tell me the export company was contracted for after news of the faked pottery scandal with the competing company, Berrocal, Ltd., broke. The interior of the entry area of Pachas's home is filled with hundreds of identical ceramics pieces that are being prepared for shipment, their curved forms stacked carefully over each other, and covered with large, protective canvas tarps. Pachas apologizes for the state of the domestic space, and the overflow of workshop products that now pack the foyer, and then invites me, graciously, to take a seat in the adjacent living room. As I weave my way through the maze of splayed and draped canvas tarps, I can't help but recall de la Fuente's unchecked, glowing descriptions of Pachas as one of the artisans she and her associates most enjoy working with, a true "gentleman" whose reserve and tact are notable among the town's other producers.

The forty-four-year-old ceramicist is by many measures among the most successful working in Chulucanas. His workshop, considered to be one of the town's most productive, is distinguished as one of only a dozen or so that's been able to achieve the ceramics output levels the largest export orders now arriving in the town demand. Along with the other workshops that are prepared to produce in large, serialized quantities, Pachas's workshops helps to complete production for orders that—reaching numbers today of seventy-five thousand items of ceramic—are without question the largest the town has ever seen. As important, Pachas has been able to maintain relationships with both of the international export companies that contract production for Chulucanas's largest shipments to US retail chains. It's this that has allowed his workshop to be one of the few in Chulucanas that can maintain year-round production cycles, giving it one of the most consistent records of guaranteeing workers steady employment.

Pachas's success is all the more exceptional—and perhaps startling—because neither he nor any family member hails originally from Chulucanas. He was one of the several skilled laborers who had mastered the technique of throwing clay using an electric potter's wheel and who the export company ALLPA transplanted from remote Peruvian provinces to introduce the technique to Chulucanas. Pachas himself was born and raised in the province of Cuzco, a region in Peru's south-central Andes mountains best known for the Incan sanctuary, Machu Picchu. Although his family had also produced ceramics that sold in Cuzco's local tourist markets, he became a student of Chulucanas's artisanal techniques only when he relocated from the mountains and arrived in the coastal town more than a decade ago. He still speaks fondly of the Andes, reminiscing about the local climate and customs of the sierra that make it regionally distinct from life in Chulucanas's low, coastal desserts. Remembering the traditions of the mountains, he tells me, speaking in soft tones, "In Cuzco, we don't celebrate Christmas in the streets, but rather with the family. But here [in Chulucanas], no. Here, they go out to dance. . . . There are differences. But man is able to change in whatever moment."

When he initially resettled in Chulucanas, he moved alone, leaving his wife and two infant children behind in the mountains, more than a day's long journey by ground. Contracted to work as part of the production team for ALLPA's first large export to the United States, he was uncertain how long he would stay, if his "services" would be needed beyond the initial order, or if the move would be worth the risk of relocating. He quickly discovered, as well, that he was at the center of local artisans' struggle to maintain traditional paddling techniques and their collective opposition to the reforms that had brought serialized production to Peru's northern coasts for the first time. Remembering the local divisions that began with his arrival, he tells me that it was the initial period in Chulucanas that were the most difficult for him: "There was a certain resistance [to me], there wasn't amicability, friendship. . . . In the street, I wasn't well-seen. The other artisans didn't look upon me well. I didn't go out a lot. . . . But with time, things would change."

Despite his central role in initiating the radical transformations in ceramics production in Chulucanas, Pachas speaks earnestly about his deep respect for the traditional techniques that defined Chulucanas ceramics. Although he insists that importing the electric potter's wheel was a change that brought "progress" to Chulucanas and helped it "evolve," he speaks with reverence for the traditional techniques that local artisans had fought to preserve. He still remembers the first time he watched local artisans at work paddle molding, recalling the feeling of awe it unexpectedly inspired. Years later, he is still convinced that it is part of what connects artisans intimately to their craft as a form of creative, personal expression: "There *is* something more magical about it. . . . When I arrived in Chulucanas, and saw the paddle, and saw the finish of the ceramics, that's when I gave it value as an art. And that [was when] art was born in me, too. [What I did] in Cuzco was something more commercial, for money, more economic than an art. . . . It's in Chulucanas's ceramics [that] I began to feel emotionally a feeling of completeness. . . . How can I explain it? Like a realized person."

Pachas's reverent description of Chulucanas's traditional techniques and his feelings of personal indebtedness to the craft are one of the rare instances that I hear the owner of a large workshop express such genuine, unrestrained devotion for the town's artisanal traditions. His descriptions are all the more striking to me—not only because he was in fact part of the force that brought about the displacement of Chulucanas's traditional techniques or because he was an outsider reviled for advancing such a displacement—but also because of a rumor circulating between local producers that I hear while visiting several workshops in the course of attending the weeklong MINCETUR-sponsored celebrations in Chulucanas.

The gossip circulating around town had it that the recent ceramic counterfeit scandal involved only the handful of modernized workshops that were able to produce the large quantities export orders required. And word was that Pachas's workshop had been among them. He makes no explicit denial or confirmation of the allegations around his workshop's involvement in the fraudulent shipment and the faking of black tint on the pottery pieces and instead explains to me how contemporary production conditions could allow something like it to happen. He describes the measured, time-intensive process of traditionally darkening Chulucanas ceramics through repeated smokings in clay ovens, typically heated by burning locally gathered mango leaves. With each smoking, resin from burnt leaves steadily settles onto the ceramic's exterior, allowing a pale surface to gradually darken from yellow and amber shades, to tones of brown and black. Often requiring up to eight separate smokings and coolings before a rich black tone is acquired, the process is slow, and moreover typically yields slight variations in coloration depending on where resin falls and sticks.

Neither the labor required to traditionally darken ceramics nor the unique pieces that are produced through the process, however, are given much value by international exporters. He tells me, "This [black color] comes from a process of natural smoking. It's not always a thing that comes out perfectly. But the [export] business demands that it's perfect. . . . And when we are fabricating 1,000 pieces, all alike with the same tone, there are going to be differences. Just as in that [misproduced] order, there were. But [the exporter] didn't accept these differences. He wanted them all the same. But this is naturally produced, where the smoke has an effect that's created inside the oven and that one can't control. And this he didn't understand."

Pachas says he's been encouraged by export companies' personnel before to "throw on a little more black paint" to quickly produce a consistent, uniform black surface between individual ceramic pieces. When he explained the ceramics' pieces black surfaces were acquired through a gradual process of smoking and not through an instant application of paint, he was nonetheless told that "something would have to be faked" or the shipment would risk getting rejected by the US chain it was contracted by. It's clear from artisans' accounts, however, that although such disregard for traditional techniques isn't necessarily always as brazen and cavalier, traditional production has nonetheless been incrementally displaced as Chulucanas's export markets expanded. Pachas himself tells me that he laments seeing the quality of ceramics deteriorate over time, as achieving "price" reductions and maximized output has come to be valued as new measures of competitiveness for workshops. He explains how product quality has necessarily been compromised with the arrival of the accelerated production for global markets. Although export orders have grown exponentially larger, workshops are given the same amount of time to fulfill and ship product orders as in the past when orders were much smaller: "It's a question of [market] prices and demand, which each time give us less and less time [for production]. This ceramic one can't do in a little time, because it all requires a process." He continues, suddenly raising his voice: "It's like when a person who doesn't know says that this color comes from yellow paint, when it's not that! It's smoke! They go on thinking that this black is paint that has to be painted on, and it's not that, it's smoke!"

It's precisely this exporters' lack of appreciation for the labor and technique invested in traditional production rather than the suggestion that such techniques were "faked" that seems to offend Pachas most. Although for the better part of our conversation his voice maintains a certain quiet reserve, it becomes notably pitched when he speaks about what he sees as the export market's devaluation of artisans' traditional techniques and how it's affected—almost contagiously—artisans' self-perception and relation to their craft. He returns several times to descriptions of the native traditions that he encountered when he first arrived in the town and becomes visibly troubled when he describes a loss of pride among producers who once "challenged themselves" through their craft making. Such a depreciation among artisans and the erosion of their investment in and personal connection to craft making is in part something he sees cultivated through an actual devaluation of labor: "If in my time I threw 100 pieces daily [using the potter's wheel], now the throwers do 200 pieces daily. . . . If we paid three soles for this piece to be [burnished by hand], then now we pay one sol. So the quality isn't maintained. . . . [But] the export companies see that they can produce 200 pieces in a single workshop and then the prices [received for a single piece] go down. It's a lot, and it shouldn't be this way. They should keep production at 100 pieces, maintain prices, and maintain quality. . . . [Otherwise] we end up *working against ourselves*. In the future, we won't arrive at a good ending."

He speaks as well of the growing "unfulfilled obligations" of workshop owners like himself who have accommodated these changes as laborers work under intensified production times but with lower wages earned per piece and with fewer securities. Talk of underpaid labor, accelerated production times, and even sweatshop conditions indeed circulate widely in Chulucanas as conditions endured for large export orders. Of all the large workshop owners I speak with who contract teams of local producers, however, Pachas is one of the few who acknowledges any responsibility for allowing such conditions to become normalized features of ceramics production. And he is perhaps the only one who speaks directly of his "obligations" to his workers and who appears to genuinely empathize with the compounding difficulties they face in labor: "The people who work with me now feel dissatisfied and they don't work with the same drive or interest as before. . . . They don't have vacations, insurance, and what happens if there aren't any orders? There's no benefit. We feel bad. Me especially, I feel bad because I can't fulfill my obligations to them."

It's not until nearly the end of our conversation that Pachas tells me he is considering closing down his workshop, permanently. It strikes me that Pachas's consideration comes for the first time in more than a decade of production in the town and after not only having persevered local scorn and exclusion for importing the potter's wheel but also having built a record as one of its most successful workshop owners. Rumors of his workshop's involvement with the recent faked shipment scandal didn't stop other export companies to seek him out for their orders. And his children, who will be the first in his family to earn college degrees, will represent a unique population from the town educated through the fortunes of the ceramic industry. If Pachas enjoys any of this or can indulge in the spoils of his successes and his modernizing "contributions" to Chulucanas, however, he seems to do so only privately. Such pleasure or even the slightest hint of self-satisfaction, at least, remains masked to me. When he last speaks, it is with the same quiet resignation that he maintained for most of our conversation, with the voice of a man defeated—or perhaps, better said, of a man wracked with unconscionable guilt: "One can't continue like this. It's not just one [exporter], it's all the exporters. They should see that instead of selling big volumes . . . they should maintain quality and try also to maintain the prices. So that we small businessmen can fulfill our obligations to our workers. Lamentably, one can't. It's just not possible."

On Innovation and Adulteration

Thirty-three-year-old Carlos Flores was just a young boy when Chulucanas ceramics began developing a stable, international market. Sold then only as unique, handmade pieces, they began to find their way onto the shelves of a select number of art boutiques in Europe and the United States, selling in small but sustained numbers. His uncle, Eduardo Flores, was among the group of artisans who had learned how to gradually give form to a piece of a clay, molding the earth in slow, progressive steps using a paddle and stone, when he, too, was just a boy. He had picked up his craft in the workshops of local potters, who used their techniques to fashion clay vessels used for cooking or brewing the fermented, cornmeal-based traditional beverage called *chicha*. It was these utility vessels, used for preparing and storing food and drink in customs that originated with the pre-Colombian civilizations who settled in the region thousands of years ago, whose exchange had maintained local ceramics markets for generations. It wasn't until a nun from the United States, Gloria Joyce, who was working in the Chulucanas diocese in the 1960s, encouraged potters to attend to the clay sculptures and decorative pieces they fashioned but had never attempted to sell that a larger market for their creative pieces was realized, too.

By the time Flores was in school, artisans from the community had begun to discover a limited, but nonetheless steady, international demand for their pieces. Some of these artisans, including Flores's uncle Eduardo, would receive individual requests to supply pieces for foreign shops selling ceramics or would occasionally be asked to travel abroad to attend art fairs and expositions. Flores remembers spending his afternoons after school in his uncle's workshop, watching him prepare his pieces, and helping him pack them carefully before a trip abroad. Often, his siblings, cousins, or classmates would gather there, too, experimenting with stray pieces of clay under the guidance and instruction of the older artisan and gradually picking up the techniques of the craft themselves: "As kids we would go over to see him, just to see what he was doing, and [to feel] as if we were sort of ceramicists ourselves. We felt proud . . . because we really saw how the artisans of that time traveled outside of the country, how they were like ambassadors of our town to the foreign world. Being a ceramicist was an honor because it meant being a representative of your town."

On the afternoon that we speak together, we are seated on two benches under the shaded portion of his workshop, a sun-washed, open area that extends from the rear of the family's home. As we speak, Chulucanas's thick late-day sun rises above us, heating the air and sending bright, flashing sheets of light across the ground. A collection of recently shaped potteries dry close by under one patch of white light. Just in front of them, Flores's young son plays sitting barefoot on the ground, his toddler's gaze turned downward and his small hands rolling pieces of clay between them. Though barely five years old, his fingers already recognize the right consistency of malleable clay and ably fashions his medium into the form of a farmer seated atop a donkey. The older Flores grins down at his son, visibly pleased by the boy's idle display of dexterity and how nimbly his imagination captures the detailed likeness of characters from Chulucanas's daily world—in all its small minutiae—into the damp earthy pieces. He asks me later to take several photos of his son posing with his creation, and I notice then how the boy had even completed his farmer figurine with a clay version of the wide-brimmed straw hat that local men wear into Chulucanas's sun-drenched fields.

Flores, too, holds one of his own unfinished clay sculptures between his hands during our conversation. He handled the ceramic body carefully as he began to speak to me, shifting the weight of curved figure between his palms for several minutes before setting it down. The sculpture, an abstract rendering of a horse, is one of the unique, hand-paddled pieces his workshop is known for producing. Although he occasionally works with an electric potter's wheel, helping to complete orders for larger workshops when his own orders are low, he is undoubtedly one of Chulucanas's few artisans who has been able to maintain an international client base for his hand-paddled pieces.

Well aware of the increasing rarity of artisans of his kind, he laments the privileging of serially produced ceramics that the export market created in

Chulucanas and what he calls the "tremendous decline" of local ceramics that began with the arrival of the electric potter's wheel. "Really many people no longer maintain their techniques because more emphasis has been given to the industrialized [model], to mass production, which really has taken away the value of Chulucanas's ceramics. Because now the seller no longer sees it as something that has to do with traditional culture. Rather, he sees it as something having to do with utility or decorativeness, just like any other craft, and it's not just like that. Chulucanas's ceramics has an entire context, right?" But over the several hours in which we speak, he makes no particular defense of the traditional techniques that Pachas had grieved when speaking with me. Flores instead emphasizes what he sees as a shift in how the population of the town in general, and how its youth especially, read and relate to the long-observed customs surrounding ceramics production. He keeps one palm over the curved, rounded form of his clay sculpture for much of time he talks, as if he were channeling his words through the piece, asking it to assess his testimony or perhaps play witness to him. "Before, these same kids, the youth of that time, really admired what it meant to be an artisan. What's different now is just that the children of artisans don't want to be artisans now. Because they see what really is the suffering of artisans. They see how their parents break their backs working. . . . It's a sacrifice because the only thing they manage to do is earn just enough to give their kids a little more than they had, but more than that, no."

He marveled too at the open-door policy his uncle and other artisans of the town practiced that allowed anyone with interest to learn the craft to enter their workshop floors. And he recalls the camaraderie that artisans maintained between themselves: "Back then, when [Chulucanas didn't have] this species of businessman artisan, they would all get together to have a chicha, most were all compadres. But it all began to, I don't know, they began to have rivalries between themselves. Jealousies. And I think it was a consequence of all this competition that there was between artisans." Flores credits this rivalry for the suspicion bred between town artisans and the declining openness among them. "I get along with everyone . . . but there are times when I encounter a rejection, someone who says, 'I don't want you to enter my workshop because you are going to copy my designs.' . . . And it wasn't this way before."

But Flores's narration suddenly turns direct and pointedly matter-of-fact when he begins to speak of the scandal in fraudulently produced ceramics that had shipped abroad from the town just a few months earlier. He describes the discovery as a shame-ridden event that could threaten the image of all production generated from the town. But even when explaining what lead Chulucanas's largest producers to participate in the hoax and to not only break with traditional modes of production but also to attempt to pass the faked ceramics as "traditionally" made, he defers from making any romanticizing references to native or ancestral tradition. He dryly describes how export contracts have steadily paid workshops less and less for supplying product orders and then says, "Look, this thing is the result of something [exporters produced]. They said they want it cheap! So they can't complain much because really they wanted the product to be so cheap that really [any other production] wasn't possible."

He tells me in fact that accelerated production times and reduced payment rates for international orders no longer allow workshops to preserve either the technical or quality standards that had been customary practice previously. He speaks of how poor production techniques, ranging from the sloppy to the explicitly fraudulent, have now become integrated as usual part of the local production landscape. And he insists that almost all the ceramics pieces that now ship for sale abroad as part of large export orders would previously have been discarded as flawed by artisans' previous standards. Dismissing international buyers' expectation that the demands for "optimized" production would allow a product to remain unchanged as impossibly naïve, he tells me, "You don't realize it, but you're buying a damaged product. . . . Everyone exports this same kind of [damaged] product. Look, it happened to me. The last time when I went to a [trade] fair, I tell you that for me, no piece sufficed. But because the fair wanted the pieces, no matter what, they took them all. And all the pieces were damaged!"

If he speaks cynically of buyers' insistence on receiving "quality" products despite the increased pressures of production, however, a surprising idealism returns to his voice again when he describes artisans' relationship to their craft. And, indeed, if there is something that he seems to be unsettled by, it is precisely the notion that artisans would themselves begin to internalize exporters' conditions for production as usual. He recalls how artisans, disturbed by exporters' insistence that flawed ceramics ship from their workshop, would nonetheless maintain their own quality standards and would independently invest the time needed to correct errors. He worries, however, that these independent practices of artisans and the affront exporters' standards once represented to their sensibilities are gradually disappearing. And he fears that artisans themselves may increasingly see degraded production techniques as a "legitimate" or at least "acceptable" form of craft making. "Before it wasn't this way. Before, we would return pieces to the oven for resmoking [if needed]. But now, no, now you can't waste time, they say. We notice the difference, but the client doesn't. They don't even realize that it's damaged. And this shouldn't be. . . . [Because] you're getting the artisan used to selling you a damaged product, a product of poor quality. And you're getting the artisan used to the notion that this damaged piece is part of *his* culture."

Flores admits his own perception of exporters' product standards, geared as they were for the global market and international consumers' urban tastes, was once considerably more favorable. He still recalls exporters' repeated insistence that local artisans be more mindful and vigilant of the state of their products and how convincing such calls were in persuading him that exporters' standards were more discriminating than those rural artisans held themselves to. Echoed consistently not only by international exporters' workshop reform projects but also by the calls for technological modernization advanced by state officials working in CITE, Flores readily accepted that if there were any deficiencies to be found in product quality that artisans' workshops and the modest, simple tools they applied would surely be their source. He remembers, too, when his own idealizing faith in the international export market's heightened standards was broken: "I thought that since they exported so much, and they spoke so much about quality control, that to export you needed a top-notch quality control. I thought that to be able to export, I would have to make a product and then verify it to the max, but it turns out that between the artisans and the exporters, it was us artisans that respected [quality] most."

He sounds almost surprised that he could have once believed exporters' messages about the global market's exacting standards and the associated demands for rural improvement. He says that in private, exporters have admitted that export standards are surpassed by those traditionally applied. And although such traditional standards must be explicitly displaced in the process of accommodating global markets, exporters nonetheless continue to market a public image of incorporating traditional techniques into their products. "It was [artisans' traditional] work that they respected for its quality. But we thought it was the other way around. We thought that their export product was so much better than ours [because] the businessmen always told us that we should check our techniques. . . . [Even if] they have always sold the image of the paddle to export."

However much Flores has come to cynically expect an inconsistency between the public image of cultural revitalization surrounding export markets and the unseen cultural adulteration that is locally produced in accommodating them, he still becomes unsettled when he speaks of increasing local acceptance—and arguably, normalization—of such contradictions. I wonder, though, how much easier it would be if he simply allowed himself to internalize the new logics of the global market and the rapidly paced, ever-mutating information economy to allow himself to be remade to its liking without any regret for the cost of such transformations or for what was lost—and to allow himself to revel instead in what gets spectacularly produced in its place.

Augmenting Experience

On the other end of town, thirty-three-year-old Victor Caminos has been working around the new pressures of export production as one of the youngest owners of Chulucanas's large production workshops. Similar to other mass output workshops, his own was implicated in the recent faked pottery scandal. And similar to many workshops that have achieved that degree of production capacity, his own has become the object of rumors of sweatshop conditions that circulate in the town. Unlike other workshop owners I spoke to, however, he betrays no sense of having been affected by the news of the recent scandal or the accusations levied against him and is pointedly upbeat when I visit his home.

And there's little wonder why. Despite the damage to his professional reputation that the scandal produced, his workshop continues to receive orders for large shipments. And despite the dishonor that such a scandal threatened to bring to an artisan and the risk of ruin it could bring to a career, Caminos nonetheless maintains active partnerships with both of the international export companies who supply the biggest orders to foreign retail chains. Far from suffering any dishonor from the various accusations, moreover, Caminos's workshop had been recently distinguished as one of only six qualified and advanced by the state to be granted the rights to use the DO. Once granted approval, he would be among the first local artisans to have his workshop's products authorized by the IP title's mark and esteemed with the global visibility its distinction offers. He's aware of the talk that circulates around the town about him and other large producers but he's casually dismissive of the accusations levied against them. He explains to me that artisans have complained about the export-oriented reforms' impact on local customs since exports began. But he says that the increased visibility of the town, the growing numbers of residents who are now involved in ceramics, and the new scales of craft production that were never before possible are uncontestable. "I see it this way: Chulucanas has evolved a lot economically because of crafts and if there hadn't been orders of that magnitude, there wouldn't have been work for so many people. From all these orders the majority of people of Chulucanas have benefited."

He mentions his early willingness to work with the electric potter's wheel and the criticism he received from neighbors and other local artisans, describing it as a dynamic that could not have unfolded in "any other place": "When [the potter's wheel] arrived, it seemed like it would leave you without work, [that it'd be] better with the paddle. Because it's more ancestral, more technical. But look at where we are now!" He acknowledges that artisans had "earned well" when they practiced traditional production techniques but that their sales were limited to "small quantities." And he says that the new production conditions and the expanded numbers of producers who are included in it challenge traditional artisans to continue to improve their techniques. "I think there is a very good relation between [traditional] and modern artisans [who are the ones] that want to keep growing, to grow beyond the masters, the initiators. This keeps the competition gong, and ensures that each time there are more people to compete."

It's these conditions that Caminos credits for pushing him to develop himself as a modern artisan and workshop owner and that he credits for urging him to continually expand his own limits of productivity. Unlike Chulucanas's older generation of artisans, Caminos picked up his craft not through family relations or through domestic production in the home but from helping to produce export orders in Chulucanas's first modernized workshop over a decade ago. He confesses that he had never imagined that he would become invested in ceramics production, and tells me he was unemployed at the time. The recent passing of his father and his new responsibility as the primary breadwinner for his family, however, made him expand his consideration for employment options. Through this casualty of events, he quickly became part of Chulucanas's first generation of producers who learned their craft in workshops engineered for serialized production.

After leveraging the export business contacts introduced to him through that workshop, he independently opened his own several years later, which turned him into a competitor of the older artisan who had first employed him. He remembers the first time his own workshop, then employing twenty-five people, was contracted for a large order of six thousand ceramics that they needed to supply in five months: "As they say, I worked like crazy [me puse las pilas], and I was able to complete the entire order." He tells me proudly that he now works with a team that's double that size for orders that are nearly three times larger. Continuing eagerly, he says, "I'll tell about an experience. Last year we had an order for sixteen thousand pieces that we had to complete in four months. Fifty-five of us had to work, working day and night. And it was a goal we were able to complete. Sometimes, at night, I would oversee the entire team—there was always someone watching." Although he says that the breakneck production speed cost him sleep and may have compromised his physical health, he recounts the experience in exhilarating terms, sounding as if he never felt more alive, and certain it had been personally transformative for him: "It was an accomplishment that *augmented experience*. Because of this, I know I can satisfy orders of that magnitude. . . . I learned a lot from the experience. . . . We challenged ourselves working day and night—and, yes, it's possible! I even arrived at a goal of working day and night, three nights in a row without sleep. . . . Of course, I looked like a stick afterwards. [But] it was an accomplishment, a really wonderful experience."

The conviction of Caminos's narration is striking as much for his specification that producing that initial export order had provided him with an "augmented experience" as for his certainty that doing so had allowed him to discover that he was the kind of man capable of enhanced life and living the "augmented experience." It is precisely this kind of accelerated living that Caminos is convinced traditional production, with its mired customs and patiently timed techniques, would never afford him. Significant as well, in Caminos's account is his self-crediting for having managed to produce an experience of augmented life. If it was the global export market that granted the opportunity to compete for an enhanced life, its realization was only contingent on the successful delivery of goods to that market. And if the sacrifices were great—and surely growing—along the way, it was only because the reward itself, however uncertain in its realization, was too. That, at least, is what Caminos tells me he came to discover after he began to produce for exporters. Indeed, convinced that traditional production could never sustain his interest, he tells me that he never held any particular esteem for ceramics during his youth in Chulucanas. That it now provides him such heightened sensations of personal empowerment is something he still marvels at: "I never thought I would be part of [ceramics]. . . . It's because of [my achievements] that I feel able to accept any order, regardless of size. . . . It was a rhythm of life that was implanted, right? That's implanted and that one adjusts to."

Caminos's success, however, and the recognition he received not only from exporters but also from the state itself through the DO titling, is notable precisely for its casual indifference to cultural authenticity. He makes no claims to practicing or even feeling an affinity for traditional production. And, indeed, it appears he feels no need to. If Caminos's workshop has been distinguished by the state as among the first considered for entitlement to use Chulucanas's DO and have its products profit from its signification, then it is less from his ability to claim native selfhood than for a different kind of self that gets actualized. Indeed, even Caminos's exhilarated accounts of the "augmented experience" accelerated production gave him and his ready, craving anticipation for the next one betray little sense of romance around indigenous artisanal production.

The contemporary state and market that seek to authorize and authenticize "indigenous" production no longer demands that the same integrity of traditional production practices from generations past be maintained. In the logic of today's global market, difference itself can be framed as a relative perception rather than a thing that can be "truly" disingenuously or fraudulently performed. In the decentralized, hyperdispersing global market, what matters in a traditional product is that it can be "staged as native"—or in the words of de la Fuente, that it can "transmit authenticity"—and maintain the appearance of its genuineness and credibility to international consumers. That under the white lights of a retail store's floor, it can still signal a "true" connection to a primordial, ancient tradition practiced thousands of miles away and that to foreign eyes, it could pass as authentic.

The DO's legal capacity to credentialize such performances lends itself perfectly to such flexible stagings of native "authenticity," allowing serially produced, nontraditional products to be authorized alongside traditional production techniques for remote, urban, and internationalized publics. In the process, the neoliberal state emerges as the ultimate postmodern, multicultural hero. Newly evolved, it no longer clings to an antiquated notion of indigeneity as born of the impossibly primordial self and instead rewards strategic, native stagings only for the purpose of capturing global consumer investment. Allowing the market's financial rewards to be channeled into participating "native" actors, it operates with a logic of enhancing recognition—not for all "authentic" natives but for those who best demonstrate themselves as "optimized" performers in a mutable information economy.

Authentic Mutations

Back in Lima's middle-class Surco neighborhood, Sonia Cespedes, a sociologist and development consultant who had worked with Chulucanas artisans to help implement the workshop modernization project for the town's first large export order, is speaking to me about the successes of the export-oriented transformations. She spent several years shuttling between her home in Lima and the town, a day's journey by road in the northern coast of the country, to work with the artisans. A ceramics maker herself, she worked intimately with artisans to identify production "inefficiencies" in their workshops, learning and adopting many of the artisans' own design techniques in her own craft. She speaks to me at length about the stunning vibrancy of Chulucanas artisanal life today and how the town now appears to have transformed itself into a "city of artisans." Indeed, it is the same kind of narration of the democratizing market that ALLPA's Maria Carmen de la Fuente speaks of in the interview published in *El Comercio*. It's precisely the export market's exponential increasing of production that has obligated the expansion of the town's productive population and that has allowed new actors—including ones who had never been personally invested in artisanal traditions or made claim to a personal history of ceramics production—to become new beneficiaries of ceramics' newly globalized commercial demand.

In the course of our conversation, I tell the usually effusive Cespedes of the news of the scandal in faked ceramics that had erupted in the town during the holiday season. Her jaw goes slack and she is for several moments completely speechless. When she comments again she repeats her complete surprise and then disbelief, proclaiming to me that such disgraceful acts could never happen in the Chulucanas she knew, that the artisans she remembers would never have conceded to such a shameful disregard of their own traditions and adulteration of craft. When she learns that it's the largest workshops producing in the town today, owned by many of the artisans that she had worked closely with for the first export-oriented reforms, she becomes almost defensive in her disbelief, and tells me that I must certainly be mistaken.

She is well aware, however, that it's the large production, export-enabled workshops that today speak for and represent Chulucanas's official success story. And the sort of authentic regard for tradition that Cespedes nostalgically recalls has, by artisans' own accounts, long been a disappearing feature for these workshops. Indeed, it is this success story that multiple institutional actors—from MINCETUR to INDECOPI and from Lima-based export businesses to the retail transnationals to whom they ship local products—retell as testimony of their own well-placed past reform efforts and the certainty of their future successes. When MINCETUR announces the short list of workshops that they have authorized to begin the process of adopting the use the town's new DO several months after the town's scandal, it is celebrated with another state-sponsored conference and ceremony. Despite the emphasis made in such events of the IP title's protective work around Chulucanas's traditional techniques, none of the artisans named practice traditional methods principally today, and two-thirds of the six workshops named produce almost entirely for serialized, large-scale orders.

Half of the six named, in fact, were among those who were alleged to have participated in the faked ceramics order that scandalized the town.

Artisans such as Carlos Flores knows that although the scandal may have marred the larger reputation of the local producers to outsiders such as Cespedes who manage to learn of it, there has been little immediate consequence for the large producers who were said to have been involved. Even though town lore now has it that the exporter who contracted the faked shipment of ceramics traveled to the town from Lima, and arrived there to confront the artisans he had contracted the order to, asking tearfully how they could have betrayed his confidence, the large workshops not only continue to produce but also are reportedly more productive than ever. Local artisans tell me that the same large workshops who were reportedly involved with the faked order were not only saved from public shaming but also were almost instantly contracted for the most recent export order that Maria Carmen de la Fuente's company ALLPA was asked to deliver. Even the exporter who had been directly implicated in the scandalized order is continuing with his operations, if not with a business-as-usual flair, with the will certainly to recapture one. Flores tells me that he was recently approached by that disgraced exporter who, impressed by his designs, asked if he would consider partnering with him. He tells me that he couldn't help being tempted by the offer but ultimately decided against it.

Ceramics being prepared for global export shipment in a Chulucanas workshop.

3 Narrating Neoliberalism: Tales of Promiscuous Assemblage

It's not until an afternoon in late April that I finally get to speak with Julio Campos. I had seen Campos multiple times in Chulucanas, where he's lived and worked as an artisan since his youth, during my recurring visits to the town over the past half year. Several months after I first asked to speak to him about his work, and just a short time after I had left Chulucanas and returned to Lima after attending the government-sponsored festival cel- ebrating the town's new intellectual property title, I unexpectedly received a phone call from him. He tells me, to my surprise, that he is in Lima and is calling from a pay phone, standing on a street corner near Kennedy Park in the Miraflores district. Before Campos finishes telling me that he has less than an hour's free time to meet, I've already begun to quickly gather my things and I tell him to stay where he is.

Campos had been part of the first group of artisans in Chulucanas who began to actively develop an international market for the town's traditional ceramics at a time when all molding was still done exclusively by hand. He and his contemporaries were among the last of a generation of artisans that together had worked solely through the traditional, time-intensive techniques that rendered individually unique ceramics, slowly, piece by piece. In the most active days of his career, he had been one of Chuluca- nas's most prominently recognized artisans. Today, however, he'd gained more repute among local residents as one of the most visible and persistent (some would say stubborn) critics of contemporary ceramics production in the town. Until global exporters arrived there from Lima in the mid-1990s, bringing with them the new technologies that reengineered workshops for large-scale, serialized production, he was among the few artisans whose traditionally made pieces circulated and sold beyond Chulucanas's local markets. Such success, visibility, and relative renown outside Chulucanas earned this artisan deep respect within town borders, distinguishing him from other traditional potters whose product sales rarely traveled beyond local markets.

Recognized by his peers as a master of his trade, Campos had dutifully served to represent the interests of local ceramicists during the most dynamic stage of his career. For over a decade, he remained one of the leading participants in the town's first artisan collective, the Vicús Association, named for the pre-Columbian Vicús people that first brought ceramics making to Peru's northern desserts thousands of years ago. And when a crossregional network of artisans, RENAPLA, was formed in the early 1990s to promote policy supporting artisans' interests across the country, he was among the three ceramicists from Chulucanas who were chosen to represent local producers. He maintained his connection to the association even after his ceramics-making career began to wane as large-scale, modernized production increasingly displaced traditional, handmade techniques in Chulucanas's ceramics market. His role as a delegate in RENAPLA's crossregional network was, in fact, the reason he had come to Lima now. Over the next few days, the organization had drawn its members together from across the country's rural and urban zones to consider new legislative proposals for development policy targeting the more than two million artisans involved in craft production across the nation. Campos would stay in the capital city only until the meetings ended and then make the day-long journey back to Chulucanas by bus along the country's coastal roads.

When I meet him a half hour after we speak, he is standing beside the same phone from which he had just called, exactly as he said he was—just a stone's throw from Kennedy Park. Developed in the nineteenth century following Peru's independence from Spain, the park serves today as the centerpiece to the capital city's cultural and commercial shopping district. A paved stretch of space that knits together a splay of early modern public fixtures, including a cathedral, stone fountain, and a patchwork of manicured garden plots, Kennedy Park displays a striking haphazardness likely unintended by the urban, elite moderns who designed it. Ringed today by rows upon rows of upscale contemporary shops and restaurants, it's this surrounding mesh of department stores, chic galleries, tourist-friendly cafés, and traffic of passenger-seeking taxis that oddly give the park its main structural coherence. Between the mix of foreign and nationally based retail chains and their sea of Spanish- and English-language signs, a nonstop flow of international tourists and Lima's "respectable" middle-class consumers mingle frenetically. As a group of chattering shoppers passes by, I spot Campos's still frame waiting, alone, his shoulders slightly tensed and his hands wrapped securely around the handle of the bag carrying all the items he's brought from Chulucanas.

We take a table at the closest café, an establishment (similar to many others in Kennedy Park) awkwardly named in English as the Swiss Café.

I have been waiting for months to speak to Campos about his work and career and about the turn of events that would convert him from being one of the most representative participants in Chulucanas's crafts market to becoming one of its most dogged critics. I invite Campos to order something from the menu, assuring him that it is the least that I can offer him for his time. He hesitates at the invitation, however, and orders nothing more than an orange juice. His unease doesn't lift even after his order arrives and his eyes dart to the surrounding tables of diners leisurely polishing off club sandwiches and chocolate mousses in between curls of polite laughter, as if checking to see if he might recognize someone.

He keeps his hands tautly folded in his lap as he begins to speak, explaining why he decided to abandon the craft to which he had dedicated years of work: "It's just that now you don't have the will to create a piece, when someone else would [copy it and] sell it for [a fifth or a third of the price]. There came a time when one was even afraid to show someone else their pieces because they might immediately be copied. . . . Now there's no respect for the *author*. Now anyone can make this." He continues, contrasting the "authorial respect" he says was once fostered between traditional artisans with the creative denigration and pollution that modern, serialized production would bring. "It definitely wasn't this way before. Each artisan had their [design] models, their clients, and they sold. But then some bad clients arrived that began to *prostitute* ceramics, and there came a moment so *dishonorable* [*descaballada*, literally translated, 'disheveled,' ungroomed, as if just getting out of bed] of selling a ceramic at a much lower price. And this, well, brought about that now ceramics aren't valued. Values no longer exist. Only money."

Campos's tendency to not mince words and the moral intensity of those he speak are part of the reason he's gained such notoriety as an uncompromising critic of export-oriented production in Chulucanas. This was, in fact, a characteristic aspect that various pro-export actors had mentioned to me about him before, warning me to be wary of his assessments and cautioning me of an embittered disposition that left him eager for vicious rumor mongering. It's hard to reconcile this image of Campos with that of the stately, elder craftsman who helped to pioneer the international circulation of Chulucanas ceramics decades ago. And it's of no small consequence that he once held such collective esteem or that he now remains one of the only traditional craftsmen of his era who maintains a moralizing fervor against modernized, serial ceramics production.

If Campos's detractors mean to undermine his credibility, however, they've managed to embed a certain self-consciousness in him when he speaks of his relationship to specific actors. He repeats—several times—during our conversation that he's "not against" any of the varied parties that helped to develop the Chulucanas export crafts market—not the exporters, not INDECOPI, not MINCETUR or its locally based administrative center CITE, and not his fellow artisans who now collaborate for export. The explicit accusation in his tone, though muted, is still hard to completely mask. And he tells me, tempering an almost sermonlike indignation with occasional qualifications: "*We're* not against other artisans. . . . Every town has the right to develop, but within certain parameters. *We* aren't opposed to this, but there were businesses that began to export enormous quantities which made [producers] lower their prices and devalue the worth of our work in ceramics." And he leans in slightly when he begins to describe the betrayals alleged to have been committed between artisans and the reported—though unconfirmed—spread of price warring for export contracts within their own. In a hushed voice, he tells me, "As I've been told, sometimes a model [for a ceramics piece for export] is received, and it's taken to various workshops. I haven't proven this, but this is what various members have told me."

Even with his qualifications, however, he doesn't stop himself before he breaks from the tones of civil diplomacy several times in his account. And Campos makes no apology for the disgrace he does unveil or the shame portioned out in what he recounts. It's this unflinching portrait of Chulucanas's export market—one that's now courted by other local artisans— that's made Campos a controversial figure among his peers in recent years as well. And if it's a clear conviction that lends itself to his condemnation of export-oriented production, it's with an audibly contrasting uncertainty that he speaks for fellow artisans' interests and his compromised alliances among them. He turns pointedly reticent when he begins to speak of relations between Chulucanas's producers and his declared commitment to preserving their social cohesion: "I like to make constructive criticisms, not destructive ones. . . . I believe that things only get fixed by talking about them. . . . [But] now it's not exactly one person that's the villain of the film, there is a whole environment. . . . The idea used to be that we [artisans] would work together and united, but this doesn't matter now." He says this to me as the boldness and indignation recede from his voice. Listening to him, I can't help but think that it may be his own conflicted status among artisans he once considered his peers and to whom he had once devoted much of his work that partly accounts for his general public discomfort and self-consciousness—traits I notice even when I see him in Chulucanas.

It's this uneven relationship to Chulucanas's expanding and newly diversifying body of producers that he takes care to clarify before we part.

Adding that he's made his peace with the new developments in his home-
town and the tensions that now mark his relations with other artisans, he
tells me, "These are things that happen. And I don't have any bitterness
toward anyone." His eyes turn downward for a moment with a quiet grief I
thought I had only seen in those of a long-devoted lover, rejected. When he
speaks again, his gaze remains downcast and introverted, as if it no longer
mattered that he were speaking to someone or as if he were sitting alone.
"I know my people of Chulucanas, and I understand that you aren't always
going to get along with everyone."

Although I had heard much of Campos's account before, our meeting
undoubtedly leaves me with a lasting impression. Having spent the past
several months speaking with the various actors involved in developing
Chulucanas's DO, including, of course, the artisans operating the largest
exporting workshops and who represent the majority of those who are
authorized to use the IP title, descriptions of compromised community ties
are ones that I've heard before from a number of other artisans. Such tales
of contemporary uncivil actions within the community and social disgraces
between fellow artisans, of broken local loyalties and confidences in the
interest of building global ones, and a degradation of authorial respect, in
fact, had been repeated frequently to me, so frequently and in such various
forms that within several months they had unexpectedly become familiar.
When I first arrived in Chulucanas, I had taken such signs of fractured
communal relations to be a confounding curiosity. Given the town's new
global distinction for having been rewarded an IP title for its traditional
craft production and its selection for new state and private investments, I
had expected to find there a community in celebration or to find some form
of public expression of a revived collective consciousness among artisans.
From the MINCETUR and INDECOPI government officials and the international exporters I had spoken to in Lima, all I heard had been resounding
praise for the various public and private efforts that had collaborated to
ensure Chulucanas's incorporation into a contemporary market economy.
Almost uniformly, and with a nearly unhalting testimony on the integrating, salvational power of global markets for traditional producers, I had
been ensured that Chulucanas was an exemplar of rural development.

That I arrived in the town to discover a radically distinct set of narrations
was a puzzle for me. Despite the enthused reportings of rural transformation that circulated so avidly in Lima, Chulucanas's own producers evinced
little collective excitement or celebratory zeal over the transformations that
had locally unraveled. And for all the assurances I received of the civilizing

successes of international export, there were very few signs of local artisans actually coming together there to recognize, celebrate, or enjoy a simple, unhurried moment of pleasure in the results of the professed developments. Citing deadline pressures and the necessity of keeping production uninterrupted, some artisans (particularly those operating the town's largest workshops) initially refused to speak to me at length or in shared public spaces, keeping the doors of their workshops closed to town outsiders and nonworkers. When producers there did begin to speak to me, they did so privately, individually relating stories of cultural betrayal, design stealing, price warring, wage exploitation, and sweatshop production as practices that had pointedly grown among artisans in recent years.

Typically, artisans relayed these accounts to me in private: on isolated roadsides, in one-on-one conversations, or in their homes. Often waiting until we were a safe distance from the town's public spaces, they expressed general distrust in the state institutions and civil associations that failed to end such abuses. And they related these stories to me secretively, in hushed and whispered tones, in confined spaces and private conversations, as if speaking of the taboo. There was a notable lack of shared public trust, security, and collectivity. Conveyed in the form of private talk, such narrations of devolved local relations are frequently dismissed by export-oriented promoters and even by some town residents as nothing more than mere "gossip." Warning that such idle chatter amounts to simply rumor or the irrational rantings of unsuccessful producers who were unable to advance beyond their primitive market, rumor naysayers stress the need to be wary of what attempts to be passed as objective "reporting," credible evaluation, and authoritative knowledge on local developments. Even other artisans themselves, especially those responsible for large export orders, decried the increasing incidence of suspect accounting among their fellow producers. Often apologizing for the disordering effects the local "custom" for gossip may have, well-speaking artisans explain such rumor mongering as an unfortunate reaction to the professional success that gossips themselves were unable to achieve. Such malicious speech, it's quickly reasoned, is really about no more than petty, interartisan jealousy, greed, and spite.

But to focus only on the content of what's expressed through gossip and to assess it on the ability to accurately depict truth and objectivity misses an essential point about gossip's function. Despite the attempts to highlight the deauthorized illegitimacy and incredibility of the content and matter of rumor, that it must be reckoned with at all seems to reveal that there is in fact nothing "mere" about it or its social force. (That scholars studying

diverse historical and cultural contexts and working from multiple academic disciplines have encountered and grappled with rumor and tales of
the incredible seems to evidence how persistent and resilient rumors have
been as features of society.) Indeed, anthropologists, historians, and communication scholars have elaborated how rumor and gossip are instead
deeply social activities that seek to perform social ties and assert collective
values by condemning behavior departing from community norms (Darnton 1999; Gluckman 1963; Taussig 1991; White 2000). By delimiting the
boundaries of potential individual action, such talk operates to shape and
construct socially shared space. Or as Luise White (2000) writes, "Gossip
and scandal served to discipline people, both those who gossiped and those
who were gossiped about: both asserted values and defined community
standards" (58). Researchers have stressed, too, how gossip could evidence
one's history of living and learning within a community, creating bonds
and boundaries around those to whom it was spoken of and spoken to.
If an "important part of gaining membership in [a] group is to learn of its
scandals" (Gluckman 1963), being able to recognize, concur in, and share
a voice in condemnation of such scandal allowed individuals to perform
their inclusion as community members.

Campos's repeated references to a collectively shared condemnation of
new export practices and his use of *we* when speaking to me is crucial, then,
precisely for the way in which it imagines itself as speaking on behalf of and
representing the voice of other members. Summoning and imagining the
force of an existing (even if invisible) community and drawing attention to
the ravaged trust, worker exploitation, and hypercompetitiveness that now
define contemporary relations around ceramics, such narrations mean to
conserve and defend the social ties that new export-oriented transformations in the town are seen to have undermined. Crucial, too, of course, is
the function of gossip to convey not simply ordinary transgressions of community norms but also extraordinary ones. Its emergence and circulation as
a form of speech and the particular tenor of its language can be especially
linked to pitched moments of crisis, radical change, and transformation
in communities—to moments, that is, when social invasion, insecurity,
and threat were not simply imminent but already powerfully underway.
Anthropologists have noted, then, how the passing of rumor, gossip, and
tales of the incredible were especially prolific among enslaved populations during the colonial era when the constant and unpredictable threat
of violence produced vampire stories, devil sightings, and innumerable
tales of terror (Comaroff and Comaroff 2005; Taussig 1991; White 2000).

Attempting to convey psychological shock and condemnation under conditions of upheaval, such moral tales then "provided a powerful way to talk about ideas and relationships that begged description . . . [and that] were so important that they were talked about with a new language" (White 2000, 30). In the face of forces of global mutation that prove beyond local control and counter to multiple local interests, such means of ascertaining and performing local ties—and testing for what might remain immutable and constant despite conditions of rapid change—emerge with new and intensified value.

That Peruvian rural artisans such as Campos would speak through an explicitly pitched and moralizing language of promiscuity—narrating the "prostitution of ceramics," of the "disrespect" for traditional "authors," and the "dishonorable" moment that fell on Chulucanas artisans when serialized production expanded in the town—is no small matter. By narrating promiscuity, local "gossips" mean to critically highlight several things about the transformed conditions that now define craft making in their town. First, export-oriented markets promote relations in which obligation and fidelity are minimized. Loyalty is practiced to the extent that it is convenient and benefiting to an individual actor and is an act expected to be fleeting and highly contingent when expressed. And partners—as they are for export companies, for instance—can be multiple and flexible, depending on how well they meet the specific demands of a suitor. Second, such relations can be seen as promiscuous to the extent that they can be read as exceeding or in violation of community ethics and threaten notions of social trust, health, and security. Third, such relations ultimately can be seen as fostering competition between actors seeking the limited affection, attention, and favor of a suitor. In this case, competition between artisans generates conditions of (at least temporary) advantage for exporters, who may be promised reduced prices or accelerated production times to secure export contracts. And last, such relations ultimately operate to dissolve spaces of solidarity, collective confidence, and public trust. Interests are recentered around individual need and become acutely self-defined and self-serving. And similar to an illicit romance, these relations come packaged with a shroud of unconfirmed accusations against competing artisans that serve only to further erode social bonds.

And if there's a force and drama behind such accounts of promiscuity and the vivid or embellished language lent to it, it's in part because such powerful conveyance is perceived as necessary. Such forceful descriptions are issued to defend against and counteract the strength of perceived external opposition and asserted master narratives. If the dominant accounts about Chulucanas that emerge from Lima's capital city and that circulate nationally celebrate export-oriented development in rural towns and preach of the cultivation of civilized, modern relations through global markets, rural gossip's counteraccounts speak of local relations as instead morally compromised, radically individuated, and increasingly promiscuous. Insisting that the disordering capacities of the free market and not just its alleged "civilizing" potential be recognized, such narrations highlight the radical absence of shared public spaces and devolution of collectivity in Chulucanas. They assert, in other words, a deeply moral and dramatically moralizing set of accounts in an effort to preserve communal life and expose how troubled public space and civil relations have become.

Even as the neoliberal promoters of international export insist on the emancipating and "civilizing" power of global markets, local artisan accounts of social disintegration stress its destructive, "decivilizing" capacity instead. Although pro-export actors appeal dually to the rational "evidence" of the modern market's civilizing promise and a religiouslike faith in global capital's salvational potential to recruit participants, artisans decry a betrayal of community ethics and condemn the rise of intercommunal abuses to prevent such partnerings. Against neoliberal salvation stories and development's master narrative of rural advancement, then, emerge another set of hypermoralizing tales. These accounts renarrate neoliberal partners and export-based development's globally networked actors as participating in relations of promiscuity with diverse parties—including ones with oppositional interests to community development. And if they fail in undoing a master narrative entirely, they nonetheless serve to destabilize its legitimacy and unsettle its authority, providing a constant and inconvenient reminder of what is sacrificed in the wake of neoliberal progress and its assemblage of a contemporary, free-market-oriented network.

Such tales of promiscuity, though, cannot be read as entirely heroic. However much they claim to represent or defend the interests and stability of the "original" community, the telling of such tales still emerge as explicit responses to the rapid changes brought on through global integration. And even though such local transformations are perceived to be imposed externally, it is still actors within the community that promiscuity's narratives primarily act upon (and punish). In their assertion, then, they can effect their own set of changes to intercommunal relations, pronouncing and polarizing already compromised local social divides and producing new moral narratives of insiders and aliens, natives and foreigners, and purity and promiscuity (Comaroff and Comaroff 2005; Nugent 1997, 2001).

Still, before those "origin" stories, a first set was told that begs retelling.

Civilization and the Official Story

In the Lima-based business office of ALLPA, Nelly Canepa and Maria Carmen de la Fuente are attempting to explain to me their use of the word *creole*. The two female executives behind ALLPA have used the term to describe some of the artisans of Chulucanas whom they've come to know over the years through their work there, and they evidently mean it to capture something more specific than the word's common, contemporary application in Peru. Used in everyday terminology such as *musica criolla* or *comida criolla*, the term is most frequently applied by Peruvians today to refer to a range of nationally identified cultural forms—encompassing dances, musical traditions, and culinary arts—that had been in practice when independence was won from Spain. Emphasizing the cultural innovation (rather than oppression) that the colonial-era's intermixing of Europeans, African slaves, and native populations had allowed, the term today typically attempts to capture and celebrate Peruvians' own independent national identity as a people of diverse races and cultures, whose shared heritage was drawn from the storied many. As a phrase that technically refers to Spanish-blooded population born in Peru during the colonial era, however, it raises a multiplicity of associations—from notions of a colonial-era European elite, to the American-born Europeans who would eventually betray the Spanish crown in the nineteenth-century independence movements, to notions of the promise of cultural pluralism and the national integration. Indeed, it is nineteenth-century national liberators' unfulfilled project of national unification and promise to create a single body of citizens with common interests and equal rights that continues to haunt Peru, similar to many other American nations, in the twenty-first century. And such a broken promise is captured in the ambivalence, confusion, and even controversy that can simultaneously be provoked via the word *creole* today. It is a term, in other words, whose meaning remains far from settled and that almost always invites explanation, debate, and pointed reflection (de la Cadena 2000; Mazzotti 2008; Morana, Dussel, and Jauregui, 2008).

"*Criollito* means being . . . well, that he wants to take advantage of the situation," Canepa explains, searching for the exact meaning. "Crafty. He wants to play like he's the 'poor guy,' the victim, the guy who's defenseless," de la Fuente assists. "Right! Pretending like he's someone from the provinces who doesn't know, but the truth is that he knows quite a bit! That's what *creole* means. The word is like *el mestizo* (the 'man of mixed-blood') who learned from both cultures, and used the combination to take the most advantage of both," Canepa chimes in again. "A 'creole' *isn't a*

gentleman, it's a person that's just trying to play with you. . . . Like a game.
. . . [Like when] artisans tell us, 'I don't have enough money to return your
loan to you.' [It indicates] you're not a serious person, not someone who
keeps their word. . . . Not all [the artisans] are gentlemen. Some are pretty
crafty," de la Fuente says, sounding more resolute this time. There's a brief
pause of silence and I almost think the pair have completed their explanation before Canepa finally adds, "And it's a form of defense, too, a response
to the kind of domination they experienced before."

Although neither de la Fuente nor Canepa places any emphasis on
the historic origins of the word or its specific reference to the European-
blooded ruling classes born in the Americas, the associations with dual
loyalty across global sites, broken ties, and dishonorable behavior remain.
And at least in their characterization of twenty-first-century market dispositions, the two ALLPA partners have in a sense earned the conviction with
which they speak. Operating the leading export company that today ships
Chulucanas's largest ceramic orders to foreign retails chains such as Target
and Pier One, ALLPA is responsible for a large part of the global presence
and circulation of Chulucanas artisanal goods. Although local artisans had
already been distributing their traditionally made ceramics to a handful of
foreign sellers in the United States and Europe when the ALLPA partners
first arrived in the town over two decades ago, they sold their products
only in small, limited quantities to boutiques and galleries. Immediately
recognizing its market appeal and an untapped potential for large-scale,
international market production, de la Fuente and Canepa became the first
export-oriented team to begin to organize artisans for large-scale global production that would pave the road toward the craft's eventual IP-oriented
investments. Engineering a range of transformations and investing in various new technologies to accelerate and amplify artisans' production, they
enabled local workshops to convert from exclusively small-scale, handmade production to serialized, massified fabrication. And although dozens
of other businessmen have followed their path and established a new base
of exporting competitors in Chulucanas's artisanal market, it's significant
that ALLPA has maintained its status as a leader in the field.

To the ALLPA partners, it's also important that it was a women-led
company that forged the export market—particularly because they read
so much of what was crucial in their work in modernizing Chulucanas as
centered on having to convert its artisans from their "creole" traditions
of exchange to cultivating new "gentlemanly" forms of business relations.
Seated behind a desk, with short stacks of paper neatly arranged across its
surface, Canepa explains, "It's a whole world unto itself, the small artisanal

producer in Peru is entirely its own thing—and converting the artisan to a businessman is a complicated goal."

De la Fuente elaborates further: "What we look for with the artisans is a leader that can be this *exception*, someone with the right *mentality*. When we find this leader, well, then we push him, train him, invest in him." She stresses that what differentiates an investment-worthy artisan from the others in Chulucanas is not artisanal skill or craftsmanship but rather an appropriate pro-entrepreneurial mind-set. Such an orientation means being men of serious business who are able to recognize a market opportunity and act on it. She emphasizes that such dispositions are rare encounters among rural artisans, however skilled they may be: "He's not easy to find, it's very difficult. They may be very able with their hands, but mentally, they're all very similar, [they're] very limited. . . . In the same conditions not all react in the same way. Some know how to take better advantage of opportunities and have a more entrepreneurial mentality, a more serious mentality. Which means they make more of an effort. Because *being a man of your word*, someone *serious*, is difficult. It's much easier to say something but not do it." She adds that working with the artisans of Peru's northern coasts, where local customs—including afternoon siestas and chicha drinking—developed around a year-round desert climate, offered a particular challenge: "[Chulucanas] is a village that has the customs of the coast. And it was a lot of work to struggle against these customs, that of drinking, of meeting work deadlines. . . . Because it's another culture. But when you start a business that has to attend to the international export market, cleanliness, order, efficiency are all important values."

The women are speaking to me inside the top floor office of the building from which ALLPA runs its operations in Lima. It's an inviting, sun-drenched space that sits above the company's main showroom, where various samples of their current exportable collections are elegantly displayed, gallerylike, across dimly lit shelves and tables. When the pair first began working with artisans in Chulucanas, they spent much of their year in the town, overseeing the deployment of new production technologies and running training sessions themselves with local craftsmen in their workshops. Today, the women still make occasional visits to the dry, desert town from which the company initially built its business. The luxury of success, however, now allows them to employ an administrative team to coordinate most local operations with the local workshops they partner with today in Chulucanas. And it's in the comfortable environs of the company's Lima-based headquarters that de la Fuente and Canepa conduct their business most of the year, and where they typically meet Chulucanas artisans who travel to Lima to discuss contract details.

De la Fuente stresses to me that capturing the rules of contemporary business negotiations and being able to respectfully maintain them is a rare trait among artisans, even those of the newest generation who came of age during the export boom. "It's not just being of the younger generation [that matters], it's also mentality, attitude." She specifies that the artisan who successfully partners with their company embodies particular "modern," "honest," and "honorable" traits: "[He's] looking to self-modernize, he's open, he knows how to communicate. . . . He's someone we would call a *gentleman*. He's a person who's very formal—he'll promise you something and he *keeps his word*. And this isn't something you'll always find. Sometime they'll tell you something and then there are A-B-C reasons why they don't do it."

De la Fuente insists indeed that their work depends on encouraging a totalized transformation of rural producers and promoting their conversion into men of business who can honor and respect the norms of the contemporary economy. Such change, she stresses, is all the more radical given it entails fundamentally redefining the very terms under which rural producers understand themselves. Speaking of such an evolution of modern selfhood as a civilizing process, she explains that it involves not only embracing the identity of an entrepreneur but also entirely eliminating and undoing identification as artisans and men of craft. As de la Fuente puts it, "It's important for us that an artisan, that *he no longer be an 'artisan'* but be a businessman." She adds that one crucial way an artisan can evidence his successful transformation into a modern entrepreneur is precisely though "honoring" ALLPA's financial loans and investments. She specifies that such modern traits and values of civil exchange, although not naturally present among rural artisans, can nonetheless be fostered within them through participation in the contemporary market economy. "We don't believe that it's good that they receive everything as if it were a donation, because then the investments end up being money that doesn't cost them anything, that they don't value. So we believe that the businessman should have the custom of using external capital and returning it. . . . It's a way of educating them within their process of entrepreneurial growth. They should respect returning loans, paying interests. It's all a part of their development."

Stressing again how few men of "honor" and "integrity" were to be found in Chulucanas before companies such as ALLPA arrived and began to foster a new kind of global export ethics there, de la Fuente details the uncivil manner with which artisans customarily manage their workshops. She bemoans what she insists is the persistence of unethical, crude, and even brutish behavior among Chulucanas workshop owners: "They don't value the work of their people, they don't pay them well, and their workers leave. They don't develop loyalty." She adds how an underappreciation for "loyalty" among local artisans is notable also in the general absence of collective forms of expression that would mean to convey collegiality and alliance: "In Chulucanas, like in many other places, there was a lot of distrust. If one artisan advances, the others would look at him with jealousy. And if one artisan learned something, he wouldn't want to share it with a colleague or neighbor. We've always felt that there wasn't enough interest among artisans in helping their village. We feel like they're a little selfish." Explicitly rooting selfishness as native to rural artisans and suggesting how deeply local custom, habit, and nature run in them, de la Fuente insists that such self-interest has persisted despite exporters' prescriptions for more civil and civic relations: "We've always believed that they should involve themselves more in the life of their community, participate more in activities that will take Chulucanas ahead, to see in what way they can collaborate for the best benefit. . . . But there's not much of this."

She shakes her head, then, as if to emphasize how much she laments rural artisans' persistent self-interest and purportedly natural disposition against modern, civic relations—even despite exporters' alleged attempts to foster more modern mind-sets and collaboration: "They should involve themselves more in the community. . . . They're not interested in cooperation. They're stuck in their workshops, accumulating money and not thinking about how they should share more with their community."

Market Manners

The ALLPA businesswomen's emphasis on crude, unruly behavior as persistent problems in villages such as Chulucanas and in the ability of the global market to correct such uncivil relations by cultivating new ethics and supposedly more trustworthy, gentlemanly modes of relating is central to the dominant narrative that circulates about rural development. And although it is only one account of recent developments in Chulucanas, the persuasive power of its narrative, particularly outside of Chulucanas's town borders, shouldn't be underestimated. If it has been able to convince, part of its ability to do so can be explained by the very non-novelty of its account and its resonance with prior Western universalizing prescriptions. Modern nation-states had cultivated and secured varied forms of civically oriented participation from diverse peoples through the promise (or "myth," according to Blom Hansen and Stepputat 2001) of realizing a shared "civil" society (Blom Hansen and Stepputat 2005; Nugent 1997, 2001; Rama 1996).

Well prior to the democratic state-building projects of the eighteenth century, science studies scholars have noted too how notions of "civility" and "credibility" disciplined the particular social relations that extended from Baroque era courts' patronage of early scientific inquiry.

Historian of science Mario Biagioli, for instance, described how the refashioning of self and the adoption of appropriate forms of mannered behavior and "polite discourse" was central to the professional success of Galileo as a sixteenth-century courtier and natural scientist. Precisely because individual "nobility and credibility were perceived as related" (Biagioli 1993, 14), natural scientists had to be concerned with "fitting into the culture of princes, patrons, and courtiers" as a central part of their work of self-legitimation as men in the pursuit of unbiased, objective truth. Historians Steven Shapin and Simon Schaffer (1985) have written also of how cultivating a "modest," measured, and gentlemanly demeanor was central to proving oneself a trustworthy member of Europe's society of natural philosophers and a worthy contributor to the seventeenth century's community-of-knowledge workers. Emphasizing how the performance and "display of a certain sort of morality was a technique in the making of matters of fact" (65), they describe how natural philosophers of the era worked to adopt particular "moral postures and appropriate modes of speech" (66) as a means of developing civil, courtly audiences for their work and establishing their own professional credibility. Specific to such trust-making tactics was demonstrating one's self as a man of "noble" character" who cared more for the "advancement of true natural philosophy" than personal reputation, personal advancement, or "individual celebrity [which could] cloud judgment" (66). Such a breed of "sober and modest" men distinguished themselves not only from other natural philosophers but also from the larger population more generally, and could be relied on as more credible, precisely because their internal balance, civility, and judiciousness would presumably keep them from "assert[ing] more than they could prove" (65).

There is, of course, something distinct to the exporters' narrative. If it were a civilizing force that could be bestowed on privileged classes generally (and men of science, especially) it was still a power that could be exercised by a sovereign political body alone. It was the sovereign power—not a wider democratic public or a global market—that was the primary guardian of civil relations and in whose hands the rise of civilization, with its promise of order, peace, and stability, rested. Prior to the growth of modern, legally regulated, and regulatable economies, the market was perceived as counter to the civilizing forces of sovereign power and government. "Ruled" by the treachery of pirates and seditious, unruly masses, markets were imagined as popular and common spaces where objects of duplicity, deauthorization, and noncredibility could flow freely. Its space was read as an ungoverned territory that left the conduct of the unmannered unchecked, and that fed the excessive, extra-legal indulgences of the nonvirtuous (Philip 2005).

The advent of modern, liberal nation-states would allow civilization's promise to extend beyond the exclusive circles of trust constructed by noblemen and aristocrats alone to a broadly conceived population of ordinary citizens (Anderson 1983; Mallon 1994, 1995; Nugent 1997, 2001; Rama 1996). Modern liberal states, it was ensured, would in fact seek to integrate rural subjects and generalize civil behavior by extending ordered, lawful relations through the building of public institutions, schools, courts, and hospitals (Mallon 1995; Nugent 1997, 2001; Wilson 2001, 2007). Through the administration of such centers, newly formed modern states could see to the assimilation of diverse subjects and the creation of a new body of ordered citizens, thereby cultivating a common civic literacy throughout such populations. If the market had long existed as the foil to the civilizing work of the royal sovereign and the emergent modern nation-state, here it would be recovered. No longer a space that cultivated acts of sedition and that trafficked in the forces of those opposed to the work of states, the market could today become cast as the central force in the remaking of untrained, rural artisans. By global exporters' accounts, indeed, it is only through the global market's acts of local recovery that remote, rural artisans can be transformed into virtuous, credible, and modern men of honor at all, and as such, be granted the opportunity to get drawn into a civil society.

It's this that explains the ALLPA exporters' description of untrustworthy, dishonorable artisans as *creoles* and that accounts for their striking departure from the term's contemporary association with national independence in Peru. In the interest of displacing the modern government as the central source of civil relations and asserting the (global) market in its place, the exporters return to a moment when creole "civility" was in pointed transition. Biologically linked to the Spanish metropole but physically and culturally removed from it, the locally empowered population of creole landowners historically posed a consistent challenge for colonial rulers, being at once "crucial to the sovereign's power and a menace to it" (Anderson 1983, 59). They were, in other words, a population whose loyalty to the Spanish crown was ever necessary for colonial administration but whose physical, cultural, and political distance put such loyalty into permanent question (Mazzotti 2008).

And, indeed, as founders of nineteenth-century emancipation movements, creole elites would later rise up against the Spanish crown and realize the metropole's greatest fears of rebellion before inheriting Latin America's independent nations as a liberal, governing class. Promising to bring forth a new national community by ruling with liberalism's all-inclusive hand, they promoted a universal, postindependence "mestizo" identity to unify heterogeneous populations fragmented between aristocratic and slave classes, rural and urban divides, and diverse racial lineages (Alonso 2005; de la Cadena 2000; Mallon 1994, 1995; Poole 1997; Stepputat 2005). The liberal origins of the term *mestizo* with its assertion of and ambition toward civic commonality, then, meant to break from the biologically determined hierarchies and essentialized social divides Spanish colonial rule depended on—however much such visions, of course, came to be incompletely realized.

That the ALLPA partners draw such pointed emphasis to the pre-independence origins of "creole" populations—when the term was marked by ambiguous identification, uncertain loyalty, and potential for betrayal to the Spanish crown—is significant. It recalls the colonial-era vantage of Spanish rulers and its reading of creole hybrid identities as threatening and effectively disrupting the particular vision of "civilization" it sought to order. Theirs was a shift that sought to emphasize the disruption to civilization's realization in Peru and that would credit such a disruption to creole rebellion leaders themselves. Their narration, therefore, delinks *creole* from its association with Latin America's nineteenth-century liberators and their claims to cultivating postcolonial "civilizations" through establishing modern liberal nation-states. Highlighting creoles' status as obsequious but disloyal and double-faced members of the Spanish colony, the exporters' narration insists that "creole" and "mestizo" practices of dishonor and dishonesty were in fact persistent, contemporary problems. From their account, the "creole" rebellion had not only failed to realize Peruvian "civilization" in the immediate period following Spanish colonial rule but continued to be the main obstacle to a generalized, contemporary emergence of civilization in the twenty-first century. Their account, however, must detach *creole* from the framings historians and postcolonial scholars underscore of the unrealized promises of creole liberal elites. Such elites, historians stress, won broad support from provincial communities in the fight for independence based on promises to extend equality to all citizens of a new republic but who reproduced the same all-too-familiar hierarchies of race and class instead (Quijano 1993, 2007).

Indeed, that promoters of global export can present the contemporary market as capable of assuming the role of "civilization's" keeper now owes something to the incompleteness of such a project—and promise—in the hands of modern nation-states. If the building of modern states and the creation of a national body of citizens who were unified in their shared equal rights and access to government had depended on the ideal of universal, democratic inclusion, such a promise remains, of course, incompletely realized centuries later (Mallon 1995; Nugent 1997, 2001; Steputtat 2005). It had been an expressed pledge to secure the ideals of emerging liberal democracies—of individual liberties, citizens' rights, and freedom from the unchecked power of government—that allowed the newly independent Peruvian state to contrast its government against that of the Spanish crown and had enabled it to extend its rule, incorporate new territories, and unify distributed and dispersed populations—including numerous rural and indigenous peoples—who had formerly never imagined themselves as part of a shared community of "nation" (Anderson, 1983). It remains, however, the contingency of the modern liberal nation-state, the unevenness by which its securities, provisions, and protections were portioned out, and the very exclusivity—rather than inclusiveness—of civilization's embrace that haunts modern, postindependent statehood in Peru today.

Such a scarcity and the starkness of the unfinished project of nation building is experienced all the more acutely in rural communities such as Chulucanas. And although the international export market can now assume credit for the work of rural development and global integration, it's significant that it does so with a radically different logic and vision of "civilization" than what modern liberal states would apply. Global markets in Chulucanas, however accelerated their speed of reaching of their distributive capacity, after all, made no effort for the general inclusion of all citizens or pretense of aiming to include all local producers. Artisans may compete for the contemporary market's favor and court exporters' patronage, even remaking themselves into twenty-first-century information workers and IP producers in order to do so, but the market's civilizing embrace, when and if it comes, is conditional, partial, and always easily undone. Civilization's scarcity and the negation of civic resources and securities no longer need evidence of the denial of rights or the marginalization of populations but can come to be seen instead as completely natural. If women, the brown masses, and men of unruly constitution were universally denied the privileges, graces, and protections afforded by royal courts' crowns, it had been an exclusion based on the inadequacies of their own inevitably hopeless, fixed (rather than failed) biologically determined natures, essential character, and internal dispositions, not on the limitations of "civilized" space—however bounded and contained it may have been (Haraway 1997).

The rationalization of the inevitability of civilization's scarcity is a crucial discursive development to modernization under contemporary neoliberal conditions. It's precisely such a logic that enables de la Fuente and fellow exporters to continue to operate as they do, without apology. Her flat exclusion of artisans who do not have the "right mentality" and her professed investment in only those "exceptional" artisans who effectively demonstrate their entrepreneurial (rather than artisanal) abilities can be spoken of as not merely justified but indeed entirely necessary. Under a contemporary narrative of neoliberalism's civilizing capacity, there is no room for mourning those denied its—now admittedly—limited provisions. More than being simply naturalized, such a scarcity can be turned into its own ideal. Renarrated now as an efficiency-producing force that ensures the "just" distribution of provisions to the rightly deserving, and the denial of such provisions to the aptly undeserving, noncredible, irrational, or ungentlemanly, civilization's rationalizing scarcity in the contemporary neoliberal theater figures in as a right-making feature to which social evolution can now be credited. Recovered as precisely the motivating force behind individual improvement and advancement, it holds the tragedy of civilization's scarcity at bay, inverting it, and turning what may have been cause for grief or indignation into romantic potential instead.

That not all actors would be convinced by the romantic claims written into a neoliberal narrative of civilization, however empowered by its authors or stunning its performance, should be no surprise. Such a narrow calculus for distributing care—as the striking visibility of contemporary logics of abandonment (Biehl 2005) demonstrate—grow harder and harder to mask, particularly for those, such as many of the nation's rural poor, already intimately familiar with life at its fringes. Historians of science readily remind us how even the voluntary participation of Europe's most self-professedly learned natural philosophers could be secured by the patronage system of Baroque period royal courts, even with severe conditions or ruthless terms of inclusion. As Mario Biagioli writes of the patronage that Galileo and his philosophical contemporaries sought, to not participate in the system, "to not engag[e] in it, one would commit social suicide" (Biagioli 1993, 15). It would be to reject the favor that inclusion in shared spaces of civil interaction could provide—and to register such a refusal so absolutely that social invisibility and irrelevance would become certain.

Inside Chulucanas, however, it's a cautionary tale of a different sort that's woven. From the vantage of the town's rural workshops an alternative narrative to that spun by neoliberal promoters takes form. And in between the furtive glances and low whispers of artisans, another set of accounts from that of contemporary civilization and its alleged indebtedness to the global market circulates. Such accounts heed against a devastating promiscuity promoted through global export and warn of a new condition in which social suicide may be guaranteed rather than deferred through global market participation.

Gossiping Civil Perversion

There is construction underway when I arrive at the workshop of Arturo Silva to speak with him. An expansive, outdoor space that encompasses more than half a dozen ceramic ovens and includes several separated production areas for the teams of workers, Silva's workshop had already been known as Chulucanas's largest and most productive workshop before the additional construction project began. During times of high production, Silva tells me that the workshop space—more than twelve-thousand-square meters in size—can produce upward of five thousand ceramic pieces monthly and can accommodate teams of more than fifty workers who keep production operating round the clock.

His workshop's contemporary distinction among other production spaces in the town makes sense. Silva was the third-generation ceramicist who ALLPA chose to first couple with for its first large US export—an order of some twenty thousand ceramic pieces that would require daily workshop production to be accelerated and amplified tenfold. Silva's parents and grandparents, he tells me, had worked in traditional ceramics, producing pottery pieces in their home and supporting their family in part through their local sales. He had been the first in the family to begin to produce pieces for international sale after he began to learn new design techniques from other artisans—including his cousins, Julio Campos and Celestino Vera, who were, at the time, just beginning to cultivate renown as local masters of their trade. Still, ceramics production for all local craftsmen, himself included, had been exclusively small scale and based on traditional handmade techniques until then. Production occurred from inside the home, would only occasionally require the assistance of additional family members to complete, and never surpassed volumes of more than a dozen pieces per day. Completing the ALLPA contract of twenty thousand pieces—an order that needed to be produced and shipped in less than six months—would require drastic physical modifications to production conditions.

"My home was very small for an order of that size, which was why we had to rent out a [larger] space," Silva remembers, speaking to me from a far corner of his workshop from which we can observe the new expansion project underway. As we watch several large bags of concrete roll past, he tells me that the initial workshop transformations, such as the current one, had required substantial financing from the export company. "We realized that the locale where I worked was very small, so we rented another one where we could tailor all the areas for the entire process, which is where ALLPA was most needed since they could teach us how to order and distribute space well." ALLPA would also suggest, of course, that modern production technologies, including the electric potter's wheel, replace traditional, hand-based techniques to accommodate high-volume, serialized production. Silva explained the required changes: "The paddle and stone [clay molding technique were] going to take too much time. . . . With a potter's wheel, someone can make between 100 to 150 pieces per day, while a master paddler could only make ten a day. That was the huge change that occurred for the artisans of Chulucanas."

He recalls that the changes the export company required initially seemed so dramatic to local artisans that few could believe ceramics production could even be sustained through them. "It was such a novelty that no one could sleep, not even the company [ALLPA], and it was a topic of discussion *if* the order was going to be completed or not, because it was such an immense order. And also, in Chulucanas, they weren't prepared for such an immense order." Silva describes how his workshop was required to produce twenty times more volume than it had before and how laborers—who for the first time were divided into production teams and assigned a single, repetitive task—remained working in his workshop until midnight.

Silva, to my surprise, makes no mention of any fatigue that he experienced in his own introduction to export-oriented production, however. And despite the dramatic new pace and scale that global market production undoubtedly imposed on Silva's experience of ceramics making, it is the collective hostility he received from other artisans for agreeing to complete the town's first large export order at all that he recalls most distinctly. He tells me that when production for the export order began, "The community of Chulucanas thought it was going to break the tradition that making ceramics. . . . For us, it was a worry, but we realized that it was our only salvation. But all the artisans were against the potter's wheel; . . . they were used to working by paddle. I was the only one who allowed the potter's wheel to enter Chulucanas, and as a consequence, I earned the enmity of all the other artisans. But with time, artisans would get used to [industrialized techniques]." And he adds, after some thought, "And there's still a lot of respect for ceramics."

Today, Silva's workshop remains one of the mere two dozen or so in Chulucanas that is able to produce the large ceramics quantities international retail chains demand. And if his account echoes much of what exporters conveyed in narrating the civilizing, salvational power of global markets and of the reform that participation in it offers rural producers, it still differs by one crucial measure. His explicit mention of how there were social "consequences" to choosing export-granted "salvation" is important—precisely because it simultaneously entailed his alienation from the larger community of artisans. Opting for salvation and inclusion in the global market at once obligated his isolation from men he had once considered his peers and who now publicly voiced their collective opposition to him and the export-oriented production model he had welcomed into the town. If it was the "salvation" of a civilizing modern export markets that Silva sought to be bestowed on him, in other words, it would be he alone that it would save and he alone who would be welcomed into its "civilization" of the few.

Despite his distinguished role in the development of global markets for Chulucanas and the celebratory enthusiasm that such a fact commands outside of the town's borders, it's perhaps not entirely surprising that Silva remains a deeply ambivalent, contested figure inside Chulucanas. His history, similar to those of other large workshop owners, would continue to arouse local controversy long after the initial decision to self-exclude from a society of fellow artisans had been made. And as they do for most large-volume producers in the town, rumors and talk of sweatshop conditions in his workshop circulate widely among residents. Telling as well is that Silva's relations with many fellow artisans, family members, and local residents that had been strained when he first agreed to partner with global exporters remain compromised even today.

Silva admits that the rumor isn't entirely unfounded. He tells me wages of his employees haven't increased since large-scale, serialized productions began in the town, even though export orders have continually increased in size. Although current export orders are now more than three times the size of the original orders international retail chains made when export began in Chulucanas, workshop laborers still receive the same amount of pay per day—about five to seven dollars—but are expected to produce faster, more, and more intensively. Silva says that he is beholden to the market because prices for a single ceramic have exponentially decreased with the entry of large orders. Leveraging his own accusations of promiscuity against other artisans, too, he tells me that other workshops are less scrupulous than he is with wages. And he tells me that new, unprincipled competitors who have recently begun to produce for export and offer to complete orders for cheaper prices have driven down what he can demand as compensation for an order.

However palatable the story of the civilizing, modernizing power of global markets and however celebrated it may be outside of Chulucanas's rural zone, it's a story that has yet to fully convince rural artisans themselves. The critical narrations of promiscuity that persist in circulating in the town highlight precisely this, calling attention to the very ravaged social ties and compromised civil relations that have proliferated locally in the wake of global export's new rural investments. Against the officially sponsored narrative of the global market as a transformative, "civilizing" force for rural development, then, emerge local tales of the perversion of collective space and civil relations. And although such accounts are usually voiced privately today and assume an aspect of gossip, rumor, and tall-tale telling, the degree of social censure and isolation export-partnering artisans such as Silva still recall enduring demonstrates how very public and shared such critical expressions once were.

Even while opposing global markets' salvation stories and tales of radical rural transformation, however, promiscuity narratives should not be read as heroic champions of community unification, stability, or "civil" local relations. They are more symptoms of, rather than antidotes to, the social upheavals that neoliberal development models are seen to impose on rural communities. When export-development narratives prescribe and insist on the need for constant change to optimize products and producers for global markets, its counternarrative defensively asserts a conservative politic to resist the market's evolutionary drive for change. To preserve culture and insist on its authority, promiscuity narratives assert a rigid binary logic that categorizes actors as either with or against "community." Propelled by a pronounced suspicion around forces newly typed as external, foreign, or new (Comaroff and Comaroff 2005; Nugent 1997, 2001), promiscuity's narrators sustain a climate of distrust through their practices. Placing a new premium on what can be typed as traditional or internal to the community, it may even require that one continually offer proof of renewed personal allegiance to the community as a condition for acceptance into its "civil" relations. Although claiming to act in defense of communal bonds of trust, promiscuity talk contradictorily produces and amplifies market-produced suspicions between community members and can further compromise already strained relations. In the moral assertion of promiscuity and in the drive for social conservation, then, "community" itself is transformed. Its accusation is by no means one performed innocently or benignly—but is one that actively seeks out a living sacrifice as a visible example. In the name of social preservation, it turns on one of its own and looks to punish those who betray community norms, label a traitor, and vilify those who were willing to sacrifice their own neighbors, brothers, and colleagues in the pursuit of explicitly noncivil, individual profit.

However self-romanticized and self-idealized, it is still a version of the "community" that gossip narrators imagine they represent and return to when they speak their tales. That the passing of such accounts would itself be driven underground or would recede to private corners and nonpublic spaces and be reduced to whispers is a consequence that promiscuity's narrators could never have predicted themselves. Even when it is only outside of shared, public spaces that it can be expressed, such narrators still claim to speak on behalf of shared, collective interests of the community. Rather than being read as a defensive reaction for civil life (or what's become of it) today, promiscuity's narration instead invites contemporary accusations of being talk of the unruly and ungentlemanly, the jealous and deceitful, or the malicious and self-interested. Such inversions of civil life and meaning, however, are among the ironic symptoms of the persistence of civilization's universal form and its reassemblage in neoliberal times.

Back in ALLPA's main office in Lima, Maria Carmen de la Fuente and Nelly Canepa are discussing two competing workshops—separately owned by an uncle and his nephew—with whom they independently partner. The younger artisan learned his craft while working in his uncle's workshop as a youth and eventually came to manage production there. Although he's indebted to his uncle for having granted him his initial lessons in ceramics and business, the younger artisan earns his living today as a primary competitor to his uncle's workshop. And the two women lavish praise on the younger artisan. They tell me that few producers have so rapidly developed their entrepreneurial potential as he and that his talents have only grown since his decision to open his own workshop, even surpassing that of his former mentor's. Canepa explains: "[He's], let's say, more consistent [with his products]. [His uncle] knows more about ceramics, . . . but [he, the nephew] has a more modern mind-set in the management of his company.

. . . And there are personality issues. Not everyone takes advantage of the same opportunities."

De la Fuente adds, speaking also of the younger artisan: "[He] has a more *entrepreneurial* mentality. And it's very curious because he's the nephew, and he worked [under his uncle] before. But [his uncle] lost him. They had a personal conflict because [his uncle] didn't know how to manage the problem, which was really about communication. He just doesn't know how to communicate, and this is, in the end, a personal issue, an issue of personality." She doesn't dwell on or sentimentalize the strained ties between family relations or the personal compromises that successfully developing an "entrepreneurial" capacity would mean for the younger artisan. And then echoing her business partner, she adds, "Some can just take better advantage of opportunities than others."

The distinction the two women make is a crucial one. It's this emphasis on opportunity, personal advancement, and individual achievement that contemporary narratives of neoliberal development—and the newly fantasized assemblages of civilization—rest on. It's such a narrative, with its idealized world of flexible, mutable relations, and ever-present opportunities for self-advancement, that contrasts with the accusations of promiscuity rural artisans circulate. Against the realm of freely enterprising, hyperproductive individuals conjured by a neoliberal "civilization" narrative appear the censure-laden gossip and "unruly" village talk of rural artisans. Seeking to expose the relations newly threatened by such transformative efforts, promiscuity's narrators produce new counteraccounts that rebuke neoliberal apparitions as themselves "uncivil" and "de-civilizing." And in an effort to preserve cultural authority, they may even rigidify and polarize notions of "tradition," asserting hardened concepts of cultural origins or boundaries as a defense against social dilution. It's precisely such rigidity, however, that feels not just conservative and prohibitive but unreasonable and even compassionateless to many artisans who feel they must continually compete to earn a living, and who feel that it would be impossible to consistently deny market-granted opportunities to self-improve.

But in the contemporary reassemblage of "civilization," there is hardly any excess of compassion and the zones of genuine, dependable care have arguably become fewer and fewer. Or, perhaps otherwise said, the conditions of genuine compassion and dependable care have become transformed themselves. Civil inclusion may have guaranteed a degree of public, socially secured care (however limited and conditional) under the governance of modern states, but under a neoliberal calculus, the simple mattering of life itself has been put in question. Anything but social givens (between neighbors, partners, colleagues, or even family members), care and compassion become instead rewards to be competed for. Existing against a backdrop of a generalized neutrality (or indifference), they become scarce resources doled out piecemeal to those who succeed in performing optimally productive living. This, at least, is the caution artisans' tales of intercommunity exploitation, betrayal, and abuse seek to issue. It is something, likely, that even the ALLPA partners and other promoters of export-based development would confess. De la Fuente's and Canepa's narration of the older artisan's lagging competitiveness and what they explicitly refer to as his inability to develop proper "communication skills" posits a reminder of the calculus applied in competing for the export market's graces after all. Such favor, they'd prompt, is far from a thing to be taken for granted. And if successfully earning it is a credit to one's "personality," enterprise, and the proper exercise of individual "abilities," so, too, does a failure to gain such favor become an indicator of personal inadequacy. Indifference from the global market and the contemporary version of "civilization" that it now assembles become markers of a failed internal disposition, an inability to have self-developed, and the incapacity to cultivate an appropriately "civil" self, fitting for the needs of neoliberalism's contemporary.

Neoliberal narrators and promoters of export-oriented development, then, might leave us with a valuable lesson on civilization and the contemporary nature of "care." Having undergone its own transformations, they'd remind us, care today becomes a function that operates—necessarily—as a rational scarcity. But through such scarcity, they'd reason, does it not become newly empowered and invigorated as a motivating force? Doesn't the shortage of care become a condition whose very insufficiency serves as a global guarantor for heightened productivity and efficiency? A force with the capacity for reordering relations to ensure the continual evolution of "civilization" in whatever contemporary form or guise it may develop?

And in these narrators' instructive detailing, they would likely pose one final set of questions to us as an open, pending body of queries: in obligating social fidelities and an individual's continual devotion in order to secure it, hasn't care in its past conventional conception functioned as a liability to individual liberty? As an object that traditionally required strict adherence to social norms in order to gain, hasn't care operated as a constraint on personal freedom, self-development, and improvement anyway? Can it not ultimately become *the* deciding constraint on individual advancement if merely granted without condition? And wouldn't the independent progress of people be sacrificed were conventional notions of care and our sentimentalized attachments to it not undone?

This, at least, is one version of global relations and their remaking under neoliberal circuits that's maintained by export advocates in Peru. Elsewhere, however, other imaginaries around contemporary global connection were being sparked, bringing distinct notions of local agency and global reform with them alight.

II Hacking at the Periphery

At the beginning of the twenty-first century, IT worlds were being unsettled with the new emergence of an old technological practice. Free software, a form of collaboratively made computer programs that were largely distributed online free of charge, had acquired a new operating system. Although the practice had existed for decades among coding communities and had been dubbed *free software* in the 1980s by the renowned MIT-based activist hacker Richard Stallman, the release of Linux in the mid-1990s had injected a pitched vitality into geek and hacker communities worldwide. Within just a few years, IT specialists, technology journalists, and entrepreneurs in high-tech centers from Silicon Valley to Hyderabad were hedging bets on not so much if free software would make an impact on global software markets but how much of an impact it would have. From the brightly lit corridors of technology expos where the future of innovation was typically a lavish showcase for commercial investors and fellow innovators alike, free, libre, and open source software (FLOSS) was already being declared a class of "disruptive technology" (Christensen 2011)—one that promised to turn on their heads the dominant logics of software innovation and the established practices of closed, proprietary commercial development that most IT product markets relied on.

Although free software practices had existed for decades, Linux's release had so quickly established a loyal user base—and more critically had cultivated a globally distributed developer community who undertook their efforts primarily as volunteers—that by the turn of the twenty-first century, IT specialists and generalists alike were calling it the "Revolution OS" (Feller, Fitzgerald, Hissam, and Lakhani 2005; Lakhani and Wolf 2005; Moody 2002). Indeed, few practices seemed to be so effective at generating the intense enthusiasm and heightened investments of global free labor that free software participants—as highly skilled information classes, no less—so extravagantly displayed. Little wonder that at the beginning of the new millennium, software worlds and information-based cultures more generally, by many accounts, were being narrated as on the brink of radical transformation.

Such predictions were not unfounded. By the turn of the century, traditional technical practices of software programmers and geeks—coding, hacking, patching, sharing, compiling, and modifying software—were finding newer forms of expression in projects that involved more than simply the design and execution of software programs. As US ethnographers such as Gabriella Coleman (2003, 2004, 2010, 2013) and Chris Kelty (2005b, 2008) explain, the volunteer-based investments and enthusiasms associated with free software were increasingly being channeled through projects that involved a range of expanding public debates around the norms, ethics, and legalities of digital cultural practice. Just about everywhere where new questions and concerns on the politics of information, digital openness, and online content distribution emerged free software practices, such that its contemporary significance, as Kelty (2008) writes, "extends far beyond the arcane and detailed technical practices of software programmers and 'geeks.' . . . [T]he practices and ideas of Free Software have extended into new realms of life and creativity: from software to music and film to science, engineering, and education; from national politics of intellectual property to global debates about civil society; from UNIX to Mac OS X and Windows; from medical records and databases to international disease monitoring and synthetic biology; from Open Source to open access. Free Software is no longer only about software—it exemplifies a more general reorientation of power and knowledge" (2). If geek culture and hacker practices had once appeared abstracted, obscure, and separate from broader social concerns, by the turn of the century such social distinctions no longer seemed to hold.

Familiar political affairs of Western liberals—including the rights of free speech, assembly, petition, and a free press, as well as growing questions around the stability of property and especially IP law—were increasingly represented by the activities and concerns of hacker communities in digital space (Benkler 2006; Boyle 1996, 2010; Kelty 2008; Lessig 2000, 2002, 2005; Liang 2003a, 2003b; Patry 2009). The growing legal debates (by some accounts, a panic) in the United States surrounding a number of mainstream, popularly referenced online entities—from Napster and Wikipedia to YouTube, Flickr, and Craigslist, as well as emerging experimental models in open-content distribution and collaborative authorship, from Open Access science publications to Creative Commons' model of copyleft distribution in music and film (Kelty 2008)—vividly demonstrated how established practices of Western law and markets were already being rewritten, and by ordinary audiences and users themselves no less (Jenkins 2008; Lessig 2008).

Across global borders, too, liberal assumptions and the fundamental principles of modern government evidenced further trouble from international hacker practices. The growth worldwide of pirate parties with digital rights platforms (Hilgartner 2009; Lewen 2008; Li 2009; Miegel and Olsson 2008), the growing visibility and complex global coordinations of the Anonymous hacker collective in its varied, localized political actions (Coleman 2011, 2013), the international range of the WikiLeaks project with its targeting of national governments, as well the dramatic, spy-thriller-like arrests of Pirate Bay and Megaupload hackers in Sweden and New Zealand, respectively, each animated—with spectacular flair—how frequent clashes between the emerging global politics of hackers and the established politics of state governance were becoming. Alongside such transnational digital media events (Dayan and Katz 1994), as the mimetic spread of the Arab Spring protests have come to define, they came to represent the means and power of online digital networks to not only interface with the "real space" realm of established politics, but also to seek to alter such real space politics as well.

Their intersections—alternately dazzling, arresting, and surreal as they have been—symbolically dramatize just how unsettling the encounter between global hacker politics and the realm of the conventional, nationally anchored political and corporate powers can be. Global digital media events—including the occupation of Tahrir Square, the unfolding protests across the Middle East, the multiple denial-of-service attacks on corporate websites by Anonymous, and international police raids on Megaupload—called the collective attention of the public worldwide, flooding news channels with global broadcasts of local citizens and sustaining the attention of audiences as if they were the only event in the world occurring at the time. And this, not only because they were revealing something that could never have been expected but because they could reveal what the public seemed to already suspect to be a constant possibility at the core of power. State conceptions of lawful discipline, authority, and possibility, as much as hackers and digital citizens' sense of such elements, were being extravagantly dramatized for the world's eyes. Similar to a latent virus, each event locally symptomized a concealed face of politics—making visible on explicitly global scales what had been established, but unseen to all but local audiences, routine exercises of state power. And each, with their fantastic spectacles of citizen broadcast and acts of coordinated indignances still channeling through the ether, could conjure new shared dream spaces for global imaginaries—making possible in an instant what had once appeared unthinkable and redefining the terms of collective reality itself in little more than the space of a few digital clicks.

Questions surrounding the desires and dreamscapes of geeks and global hacker communities—ever-more pressing as the activities and experiments of free software advocates had begun to shape broader public debates around digital policy—only amplified following such events. Early qualitative studies on FLOSS communities had sought to uncover the personal motivations of FLOSS coders—seeking to explain why such highly skilled classes of information workers, ones who prided themselves on their skills of elegant code expression and rational, logic-bound program organization—could invest such vast amounts of highly valued time and work to projects they knew they would never be paid for (Ghosh 2005; Lakhani and Wolf 2005), how actors representing such model rational subjects could upset the rationality of economic laws of supply and demand. Surveying hundreds and sometimes thousands of volunteer-based coders and enumerating their individual responses for motivations cited, early studies focused almost exclusively on the United States and Europe, where the largest populations of FLOSS developers resided (Feller, Fitzgerald, Hissam, and Lakhani 2005). Studies such as that of Lakhani and Wolf's summarized their findings into four distinct rationales , concluding that FLOSS coders were primarily compelled because they (1) expressed enjoyment and learning as a primary motivator, (2) simply needed the code to satisfy non-work-related user needs, (3) had work-related needs and career concerns, and (4) felt an obligation to the community and believed that software should be free and open (Lakhani and Wolf 2005). Others underscored the "selfish" motivations underpinning the activities of geek participants, who emphasized the desire to learn new skills as a key driver (Ghosh, 2005). Individual motivations remained anchored in the rational, self-interested individual, and whatever role culture, social context, or local history played seemed—at least in the earliest FLOSS studies undertaken—largely irrelevant.

More recent research on hacker practices by US-based anthropologists, however, has explicitly underscored the cultural attributes surrounding participation in networked collectives, particularly when virtualized and geographically segmented forms of practice and labor are involved. Gabriella Coleman has noted, for instance, the intense forms of online and offline socialities that build the "lifeworlds" of hackers and that make themselves evident among the internationally shared identity practices of FLOSS's distributed community members (Coleman, 2010). Her analysis unpacks how shared spaces of digital practice, including FLOSS conferences, can be understood as ritual-like affairs, laden with practices of "confirmation, liberation, celebration, and especially re-enchantment where the quotidian affairs of life, work, labor, and social interactions are ritualized and thus experienced on fundamentally different terms" (Coleman 2010, 53). Citing Bakhtin, Coleman stresses hackers' collective gatherings as means of orchestrating "a celebratory condensation . . . [to] imbue their [individual] actions with new, revitalized, or ethically charged meanings [and lift] life 'out of its routine' (Bakhtin 1984, 273). . . . These are profound moments of cultural re-enchantment whereby participants build and share a heightened experience of each other" (Coleman 2010, 53). Via organized hacker spaces, common elements of the "quotidian technological lifeworld"—whether that of compiling code, laying down networks, setting up servers, or generally investigating the possibilities of technological solutions to problems encountered—can newly unfold and take on collective dimensions in emotionally charged conference settings: "What the conference foremost allows for is a 'condition of heightened inter-subjectivity' where copious instances of hacking are brought into being and social bonds between participants are made manifest, and thus felt acutely. . . . In short, for a brief moment in time, the ordinary character of their social world is ritually encased, engendering a profound appreciation and awareness of their labor, friendships, events, and objects that often go unnoticed due to their piecemeal and quotidian nature" (Coleman 2010, 56).

Such work invaluably attends to the means by which the ordinary, day-to-day experiences of FLOSS participants—the mundane investments of time and invisible care, and the habituated practices of work or play—can link into larger events and collective performances of shared communality; ones that not infrequently in the age of digital networks aspire to register globally spectacular effects. Free software's practices are explicitly emotionally heightened, voluntary investments of time, labor, and care—distributed across global scales—that translate into more than the technical production of code. And similar to other cultural stakes, they can be enacted to compel investments for audiences beyond fellow coders. Similar to other cultural stakes as well, the interests and dedicated engagements that channel though them are shaped by the local spaces and historically situated contexts that anchor parts of participants' active "lifeworlds"—however much engagements with global digital culture might take on clearly cosmopolitan dimensions. At a moment when the politics and values of hackers and free software geeks become ever-more internationally visible and dispersed, there remains a question of not only how one learns to "see" as a participant of free software's global culture—but a question, too, of what indeed one does *see* from the standpoint of situated, localized context? What, that is, is seen when one "sees" free software in all its ambiguously and multiply global, local, and porous forms?

There are familiar answers to such questions, including the stakes of information sharing, collaborative authorship, independent learning, and open and transparent infrastructures (digital and otherwise) that are commonly cited as universally shared social stakes of FLOSS and hacker communities. Indeed, for such interests to be globally circulated so that they can represent the concerns of diverse members across dispersed local sites is the standing promise of any universal logic. Such oft-repeated interests, however, turn out only to lead to further questions the more local contexts and local complexities are taken into account. How, for instance, do such shared stakes come to represent the interests of not only hackers from the global periphery and outside of recognized technological centers in equal kind as those embedded within centers but to also speak for the interests of rural communities, indigenous actors, and "minority" user groups who are often read as far outside circles actively designing digital futures?

If hacking has indeed shifted from an obscure, esoteric practice of professional or specialized technicians based largely in the West to a globally extending practice now also invested in connecting worlds, remaking social terms, and enacting forms of political critique that can even aspire to counter dominant forces of established states, then the question emerges of how it comes to be understood and expressed as such. What, indeed, have been the forces that have allowed its terms and stakes to be fought for and struggled over varied local terrains—and through interests that extend beyond concerns of the Western liberal coders who originally represented its development base? How, in other words, does the evidently global, cosmopolitan character of the FLOSS information culture and stakes in reform across distributed regional spaces come to be translated and anchored into diverse local sites, including beyond the West?

In her study of the emergence of global environmentalist ethics in Indonesia, Anna Tsing (2004) reminds her readers of how student activists and "nature lovers" she studied consistently underscored the importance of attending to how "cosmopolitan specificity" comes about—how the contours of its situated cultivation come forth and how the diverse local commitments made by particular actors irrefutably matter. However "universal" the presence of nature, "nature loving" itself is by no means an automatic practice. Indeed, like learning to "see" money (Simmel 1990) or the prospects of wealth as a potential of IP "seeing" nature must be cultivated. Such student activists underscore how the locally anchored, cultural specificities to globally shared ethics surrounding "universally" present and relevant objects like "nature"—or so too, those surrounding digital information—cannot be ignored. Tsing reminds her readers that however familiar the contours of universal logics and cultures of cosmopolitan orientation, differences—whether difficult or simple to perceive—always matter. "Cultures are always wide-ranging and situated, whether participants imagine them as global or local, modern or traditional, futuristic or backward-looking. The challenge of cultural analysis is to address both the spreading interconnections and the locatedness of culture. To study a self-conscious cosmopolitanism in all its energizing connections to the world and in all of its exotic distinctiveness models the inextricability of interconnection and location" (Tsing 2004, 122).

The lively divergences that surround local imaginaries of free software's global network remind us, too, of another crucial point: that neoliberal techniques and those of global free markets are not the only cultural logics that have been channeled through information technologies in the twenty-first century. Neither are such free market–oriented techniques the only technologies deployed in Peru in the interest of remaking worldly orientations across diverse local sites. However "inevitable" and "universal" neoliberalism's aggressive unfolding across contemporary sites and its remakings of global imaginaries have been in the wake, it too once existed as a minority logic of reform against other established powers. And it too must still work against other emergent cultural logics that vie to remake imaginaries around information technologies and their local digital worlds in the new millennium.

2 5 MAR. 2001

San Isidro, 21 de Marzo de 2002

Señor
Edgar Villanueva Nuñez
Congresista de la República
Presente.-

Estimados señores:

Primeramente, queremos agradecerle la oportunidad que nos brindó de informarle cómo venimos trabajando en el País en beneficio del sector público, siempre buscando las mejores alternativas para lograr la implementación de programas que permitan consolidar las iniciativas de modernización y transparencia del Estado. Precisamente, fruto de nuestra reunión hoy Usted conoce de nuestros avances a nivel internacional en el diseño de nuevos servicios para el ciudadano, dentro del marco de un Estado modelo que respeta y protege los derechos de autor.

Este accionar, tal como conversamos, es parte de una iniciativa mundial y hoy en día existen diversas experiencias que han permitido colaborar con programas de apoyo al Estado y a la comunidad en la adopción de la tecnología como un elemento estratégico para impactar en la calidad de vida de los ciudadanos.

A copy of the letter sent in 2001 by Juan Alberto Gonzales, the general manager of Microsoft Peru, to Congressman Edgar Villanueva to contest Proposition 1609. The letter, which was circulated online among networks of FLOSS activists, helped generate international media attention to the cause of Peruvian FLOSS activists. (For a translation, see the notes at the back of the book.)

4 Polyvocal Networks: Advocating Free Software in Latin America

In October 2005, Peru caused a buzz among international technology news outlets when it passed an unusual law that challenged the adoption of commercial software—in its standard closed, proprietary form—in government. The bill, one of the first of its kind globally, obligated "technological neutrality" in public institutions' selection of software and required FLOSS to be considered as an alternative to proprietary software when contracting software vendors. Its passage was celebrated by the Andean nation's free software community, who had long been advocating for state adoption of FLOSS. Four years earlier, a legislative proposal—dubbed Proposition 1609—was introduced in Peru's Congress to mandate FLOSS in state computers. Similar legal measures had begun elsewhere around the world, all seeking to establish alternatives to government use of closed, proprietary software. But it was Peru's legislative efforts that in the early years of the new millennium managed to uniquely channel the spirit of the global free software movement with its emphasis on technological liberty and reform, and capture international attention—and so too the investments of the United Nation's first conference on FLOSS in Latin American governments.

Much of the early publicity was spurred after Microsoft's general manager in Peru attacked the bill as a "danger" to national security and to corporate intellectual property rights in 2002. Not long after, the US ambassador to Peru issued a threatening letter that reiterated Microsoft's disapproval of the bill and warned its passage would harm US-Peru relations. The congressional sponsor of the bill, Congressman Edgar Villanueva, nonetheless staunchly defended Proposition 1609 in a twelve-page response that was later translated into more than a dozen different languages from volunteers worldwide. As Microsoft's interventions and Villanueva's response were circulated online, Peru and the congressional sponsor of its free software bill were suddenly transformed into prominently visible players in the FLOSS global movement. Or as one reporter from the English-language news site *Linux Today* prophetically described,

In the course of everyday business and politics, once in a while something truly significant happens. At such a time, letters become road maps for change and a politician from a small mountain town in Peru can become a hero to those who believe in a cause: both amongst his countrymen, and around the rest of the world. . . . Congressman Villanueva's reply [to Microsoft] . . . raised him practically to folk hero status overnight. (LeBlanc 2002)

How the letter, written by a provincial Peruvian senator from the Andean town of Andahuaylas, came to speak for the interests of hackers and information activists, not merely across Peru and Latin America but also well around the world, is the curiosity of networked digital culture that this chapter explores.

Speaking through Software

Although outside of Peru, the nation's free software legislation was celebrated for underscoring the internationally shared stakes around open technologies and their clear, universal benefits, inside Peru the legislation brought to light the particular, national contours surrounding debates around technological reform, freedom, and transparency. Indeed, in the early years of the new millennium, calls for greater institutional accountability and openness to publics that were central promises of free software in government resonated throughout the South American nation.

In the last decades of the twentieth century, increasing evidence around authoritarian practices, human rights abuses, and routine corruption by the state had been repeatedly raised as a two-decade long armed conflict with the rurally based Sendero Luminoso, or Shining Path Movement, unraveled across Peru. Much of this evidence came to light during the administration of Alberto Fujimori, who despite being democratically elected in 1990, would later be remembered for his authoritarian autogolpe, in which he suspended constitutional rule, shut down Congress, and purged the judiciary. He would also be remembered for the record of brutal treatment of national prisoners, political disappearances, and kidnappings during the conflict with Sendero—during which some sixty-nine thousand Peruvians were later estimated to have been killed, largely in rural provinces, and by the hands of armed state and guerilla forces alike. None of this, however, would keep him from running for office for a third electoral term in 2000—changing Peru's constitution in fact in order to allow him to do so—or from successfully winning that bid for office. And none of this, either, would keep him either from being able to flee the country only months after his reelection to seek exile in Japan after mounting evidence—including documented videotapes of acts of bribery and extortion by state officials—dramatically revealed the extent of public nonaccountability and corruption that permeated government.

Indeed, earlier that year, key members of Fujimori's administration—most notably, the head of Peru's central intelligence agency (the Servicio de Inteligencia Nacional, or SIN), Vladimiro Montesinos, who had attempted to flee the nation and seek asylum in Panama after his infamous "Vlady videos" came to public light—were embroiled in scandal. Montesinos, in a clear practice of incredible vanity and unfathomable entitlement, had been secretly videotaping his own acts of bribery with prominent individuals in his office. The tapes revealed his payments—in some cases of more than a million dollars a month—from congressmen and military officers to members of Peru's national press—to ensure various political favors. The practice proved to be only the beginning of what were a long list of crimes that included embezzlement, drug trafficking, gun running, vote rigging, and torture—of political prisoners, Peruvian journalists, and intelligence officers who were considered disloyal to the SIN. Among these, in fact, were multiple acts that directly involved the 2000 presidential reelection of Fujimori as the civil war with Sendero Luminoso and two decades of political violence had been coming to end. One journalist came forth with reports of having been kidnapped by secret police agents who sawed his arm to the bone to keep him from disclosing a videotape of Montesinos bribing election officials to fix the vote. Army intelligence agents also surfaced with revelations that Montesinos had ordered wire taps of leading politicians and journalists and had ordered the torture and murder of fellow agents who had leaked information to the press. And in the midst of such revelations, too, came additional reports that Montesinos had been supported by funds from the US government's Central Intelligence Agency, who reportedly channeled millions of dollars to him to fund antiterrorist policies.

The years immediately before the free software proposal emerged had presented a series of pitched reminders of how deeply entrenched practices of corruption, bribery, and outright thuggery could be among elected officials—and how tolerant international leadership could be of such blatant violations of public trust. Proposition 1609 and its subsequent letter-based exchange with Microsoft representatives emerged from a context in Peru in which undeniable evidence of state officials' compromised sense of accountability made all too clear the need for greater public transparency and accessibility. Such abuse of political privilege, indeed, only further dramatized how starkly divided the nation remained between classes of political or racial privilege (largely concentrated in urban zones) and those (largely dispersed in rural zones) who were kept politically remote, removed, and excluded.

Still, outside Peru, where many FLOSS advocates had witnessed FLOSS advance from its origins as an isolated practice of Western hackers to a globally dynamic phenomenon endorsed by transnational technology corporations and volunteer-based associations alike, the emergence of a particular free software–centered law was decried as unnecessary (Bessen 2002; Evans 2002; Hahn 2002; Stanco 2003). Adhering to a notion of user liberties that insisted individuals should be able to freely choose between proprietary or nonproprietary solutions, FLOSS advocates and policy makers in Western centers urged governments in the early 2000s to maintain instead a stance of *political* neutrality in software acquisition (Chan 2004). FLOSS's rapid transition from the margins to the mainstream of technology markets and usage in the United States and Europe, after all, seemed to occur free from government interventions. Why should it be distinct elsewhere? Advocates of such positions began to reemphasize framings of FLOSS—already popular in US information technology conferences—of FLOSS as a species of "disruptive technologies" (Christensen 2011) that would inevitably displace outdated technology (Bessen 2002). To the commercial software industry, such readings signaled the need for dramatic self-transformation to maintain dominance. For FLOSS participants in Western centers of FLOSS production, it served instead as reassurance that their current practices should advance unchanged. Meanwhile, media coverage on FLOSS legislation emphasized economic rationales for governments' adoption, framing it as such a drastically cheaper alternative to proprietary software that its capacity to ensure financial savings was surely driving government interest internationally (Dorn 2003; Festa 2001; Stocking 2003; Wired. com 2003). Whether stressing technological or economical merit, such oft-repeated framings relied on projections of technological inevitability and gradual global diffusion (Callon and Latour 1981; Latour 1987, 2005) that presumed it would only be a matter of time before the obvious, universal advantages of FLOSS came to be recognized by all classes of IT users worldwide.

A closer examination of the practices that surrounded the emergence of FLOSS legislation in Peru—at the periphery of FLOSS's Western centers of production—however, reveals that little sense of inevitability defined the activity of local FLOSS advocates. Far from presuming any steady global advancement to FLOSS's evolution, the proponents of Peru's Proposition 1609 undertook various forms of local and nonlocal work to advance and "translate" their interests within a conscientiously nationally framed context (Callon and Latour 1981; Latour 1987, 2005). As Latin American science and technology scholars, Medina, da Costa, and Holmes (2012) have underscored in stressing the utility of the "translation" models for tracing the means by which knowledge and technology move between discrete locations within and beyond the region: "[T]ranslation implies an inherent change or transformation as well as a displacement. It is reminiscent of the Italian adage *traduttore/traditore* (*translation/treason*), a reminder that one can substantially change or 'betray' the original text by moving it from one language to another. The translation model of technoscience [unlike diffusion models] . . . focuses on the way all techno-scientific objects are in constant state of flux and asserts that it is more illuminating to study these multiple moments of translation/treason than [diffusion model's] lone moments of invention/transfer."

Although many US-based FLOSS advocates would adopt a stance of "political agnosticism" (Coleman 2003, 2004) that read ties to formal politics as counter- or nonproductive to the essential work of technological development and diffusion, Peru's FLOSS advocates insisted instead on FLOSS in Latin America as necessarily engaged with practices of governance and political reform, and actively sought to build spaces of debate that might redefine relations with national political channels. If many FLOSS supporters in Western centers presumed and would continue to assert that FLOSS technologies could spread globally without government intervention based on technological merit or economic rationales, Peruvian advocates' activity around Proposition 1609 signaled a departure from such assumptions and suggested that other interests beyond FLOSS's technological development and innovation could be of primary concern to regional IT activists. Far from presuming the inevitable diffusion of new, FLOSS-based technologies from centers to peripheries, their activities stressed the contingency of local FLOSS uptake, and the need to cultivate local relations to enable it to be shaped, socially and technologically, and get "translated" across diverse interests in order to achieve regional extensions (Callon and Latour 1981; Latour 1987, 2005).

Theirs, that is, were investments oriented toward moving beyond a narrative of technological inevitability and dominant framing of FLOSS in Western centers as governed by logics of global diffusion. Latin American FLOSS advocates' activity insisted instead that a set of new queries be posed in the interest of highlighting not simply technological evolution but also historical, political, and local context as well. Foremost among these, then, was the question of what made the proposal of Peru's legislation and its emergence as a prominent site of FLOSS advocacy possible to begin with? What sociotechnical negotiations, situated forms of local work, and trans-regional relational networks were necessary for such an event to be produced? What were the diverse constellations of meaning that channeled through and accounted for participant actions across distributed sites? And what, indeed, are we to make of the very multiplicity and cacophony of such accounting acts that all simultaneously "speak for" the network?

Such actors' engagements underscored the notion that to peer beneath the surface of Peru's FLOSS movement was to reveal a network of local actors, distributed through multiple regional contexts and invested in a diversity of technical and discursive practices to advocate for political reform. Encompassing Latin American politicians, independent citizens and entrepreneurs working with FLOSS applications, local Linux and FLOSS user groups, Argentinean programmers, and Peruvian student activists, such a transnational network functioned as a polyvocal system to foster the emergence of new forms of civic expression and expanded spaces of political participation. Indeed, the force and efficacy of Peru's FLOSS advocacy network relied fundamentally on such polyvocality—which, despite its uncoordinated and decentralized nature, could generate new discursive formulations and challenges to dominant Western-centric assumptions on technological use and development.

A growing body of research on FLOSS communities has attended to how the mutual accommodation and production of plurality serve as valuable testimony to the strength of networks (Benkler 2006; Coleman and Hill 2004; Kelty 2005a, 2005b; Lin 2007; Weber 2004). Building on such work, this exploration of networks as polyvocal bodies traces out of the varied paths of actors who identify themselves as involved in advocating for the Peruvian FLOSS legislation across their diverse sites of promotion. These indeed thread through Peru, but they are linked too to other sites and resources across Latin America and with cultural imaginaries also stretching to the United States, Europe, and beyond. The activities of groups including the Peruvian Linux Users Group, the largest FLOSS-focused user group in Peru with more than fifteen hundred subscribed members, and the digital rights nonprofit Via Libre (Free Path) Foundation in Cordoba, Argentina, each leveraged global mailing lists as well as real-time instant relay chat (IRC) channel exchanges to organize and demonstrate such fluid, multi-sited commitments. Conferences that were organized as physical spaces of encounter were oriented to speak not only to other skilled technology users and coders but also aimed to translate FLOSS across simultaneously global and local interests. Such work reverberated through the more than dozen Peruvian FLOSS conferences I attended—in Lima and five other provincial towns and cities—during the summers of 2003 and 2006 alone. Some targeted local or national government actors, and others were organized with them. These ranged from large, international gatherings, such as the first UNESCO-funded conference for the use of FLOSS in Latin American and Caribbean governments held in Cuzco in August 2003, to local community-focused gatherings promoting FLOSS in rural sites and regional government, such as that held in the Andean mountain town of Andahuaylas immediately following the Cuzco summit.

A consideration of the diverse spaces that Latin American FLOSS advocacy speaks to and with reveals that central to the network's performance is not merely the extension and diffusion of FLOSS's technological artifacts but also the translation of negotiated values and practices with particular local contexts—and so, too, the resulting production of new discursive spaces, cultural meanings, and narratives that can testify to the technical and political value of open technologies. It reveals, in other words, how networks may function powerfully not just through their ability to globally coordinate technological innovativeness or unify multiple interests through a single, centralized, representative actor or spokesman, but through their very cultural innovativeness, nonpredictable discursive productivity, and varied local translations that multiply sites of accounting through peripheral extensions.

Network Spokesmen, Network Narrators

Early scholarly attention to networks in science and technology studies (STS) brought attention to how network structures enabled the generalized extension of scientific facts and technological artifacts. But STS demonstrated how network success depended less on the condition of factual or technological strength than on the skillful recruitment of consent from diverse human and nonhuman actors (Callon 1986, Latour 1987). The strength of scientific facts or technological artifacts emerged then as effects of rather than conditions for the productive activity of networks' varied actors (Callon 1986; Latour 1987, 2005). These theorizations illuminated the ways that networks could not only efficiently manage the labor of diverse actors but also channel actors' multiple interests under a single, spatially distributed and extending representative form. Central to this process was an initial stage of enrollment enacted through the effective negotiation of networked actors' interests across distinct local sites. Varied relations across space had to be drawn together such that a single network could unify diverse interests, harmonize voices, and be authorized to speak for the whole in a spokesperson-like status (Callon 1986; Callon and Latour 1981; Latour 2004a). These coordinations and the ability of local differences to be negotiated and represented through spokesperson functions thus explained how scientific facts and technological innovations could achieve broader adoption and standard acceptance among a body of other recognized "facts" and inventions. Here, the strength of networks was attributed to their capacity to stabilize technoscientific facts and function coherently through employing spokespeople in the work of coordination toward a shared end, despite the multiple local modifications necessarily negotiated across distinct sites.

Recent research on the new economic productivity of networked activity has highlighted instead networks as systems of technological innovation (Benkler 2006; Castells 1996; Lessig 2002; Stark 2009). Shifting attention away from the negotiation of relations across distributed local sites that translation models underscored, such approaches characterize networks as loosely organized bodies that leverage new information technologies to organize the labor of diverse individual parties and stressed their surprising success in generating new technological innovations. Such research has challenged established assumptions on innovation as dependent on tightly coordinated, closely managed, and hierarchically organized production units. But here, too, it is the network's capacity for coordinating innovation and generating and circulating new technologies that primarily determined how its strength and durability would be assessed.

Interviews with Peru's FLOSS advocates readily demonstrated, however, that no single, coherent explanatory account of the network extended through all parties. Neither was the dominant organizing logic most often asserted by Western FLOSS participants of technological innovation the principle interest motivating Latin American FLOSS advocacy. Regional network actors instead underscored their own independent recognition and distinction, refusing integration into a single representative logic that would collapse multiple concerns under a unified, "harmonizing" position or "consensus" (Stengers 2007). One heard instead a polyvocal assemblage of stories that—in their simultaneous dynamism and dissonance—revealed FLOSS advocacy in Peru as built on a diversity of regional and local practices that at times intersected but never entirely conformed with each other or with those emphasized in the West. Multiply positioned and independently speaking, local network actors generated dynamic bodies of value about what FLOSS could mean across distinct sites, why it should be promoted, and what could be at stake in its promotion.

That such noncoordinated practices could succeed in generating attention for Peru's FLOSS movement was an outcome that network actors offered no single, privileged explanation for. Indeed, no single rationale was distinguished in the accounts of Peru's FLOSS advocates, just as no single actor was given primary credit for the network's celebrated effects. Politicians, NGO workers, and individual coders who participated in the network instead continually referenced the work of others, distributing recognition to account for broader effects. Such a notion underscores that to study a polyvocal network calls attention to a multiplicity of contributions that extend lines of explanation and sites of local work. Indeed, in cases such as this, one had to be prepared to find that the productivity of the network depended less on the coordinations of centers and spokespersons than on a distinct multiplication of local accounting capacities and extensions at the periphery.

Letters, Legislation, and Disrupting Progress Accounts

Superficially, Congressman Edgar Villanueva appeared to be an unlikely proponent of FLOSS. He began his career in state politics as the mayor of Andahuaylas, a mountain-based town in the Apuirmac province whose population of thirty-four thousand people relies—as do most communities in the surrounding towns in the Andes—primarily on an agricultural economy. He was elected to Congress in 2001 and in his first year and a half there, his legislative sponsorship focused on regional and educational bills primarily, with little focus on technology or computing initiatives. Without question, that is, Proposition 1609 stands out in Congressman Villanueva's record as the lone legislative action specifically addressing the use of software or IT.

Proposition 1609's presentation before the Peruvian Congress in December 2001, however, left little doubt that its supporters were anything but novices in articulating their positions on the need for technological reform or in specifying its relation to political reform. Proposing the mandatory adoption of FLOSS in public offices, the bill's text specified that exceptions be made only when a developed-enough FLOSS application was unavailable. Emphasizing the contemporary legal contradictions experienced by governments in software use, it opened by stressing that states' reliance on computing in nearly all administrative activities forced governments into "a situation of dependency . . . [on] technology created [and priced] in other countries." The bill further cited the rapidity of software updates, stressing that the frequency of new software releases by corporations forced governments to choose between continually purchasing new licenses, operating with outdated software, or piracy. It also referenced a government study that estimated Peruvian government use of pirated programs at 90 percent (INEI 2002) as a result of inaccessible licensing fees and pricing schemes, and concluded that governments must find alternatives to "[break] the vicious circle of dependency."

Proposition 1609 thus framed the particular legal and economic imperatives of states in the Global South as central to the Peru's own interest in ceasing its dependence on closed, proprietary software. Moving beyond arguments for FLOSS adoption that stressed states' technological needs, Proposition 1609 asserted a technological narrative that critically implicated logics of global markets and international governance that marginalized "peripheral" states and maintained global relations of political and economic dominance. By the bill's account, dynamics of power that privileged wealthy nations' and Western industries' commercial interests could also be reproduced by decisions of Peru's government in its marginalizing of voiced concerns by local citizens, IT users, and developers. If state decisions around IT purchases and service contracts could have been previously assumed to be dry, esoteric decisions that sparked relatively little civic concern, Proposition 1609 pronounced them instead as matters citizens of the Global South had become conscious of as deeply politicized, socially expensive choices. And ones, moreover, that could reinscribe the nation and polity into new, extended cycles of Western dependence.

Within months of Proposition 1609's presentation to Congress, Microsoft, the primary software vendor for Peru, intervened. In a March 2002 letter addressed to Congressman Villanueva, Juan Alberto Gonzales, the general manager of Microsoft Peru, issued his own projection of how FLOSS would fundamentally compromise the state. Positioning FLOSS as a risk-laden technology, Gonzales foretold a swarm of devastating effects that could be unleashed under Proposition 1609. He warned that FLOSS would inflict immeasurable expenditures for technological migration, generate noncompatibility between Peru's public and private sectors, devastate corporate productivity, and hamper "the creativity of the entire Peruvian software industry" (whose IP rights would somehow be compromised). Arguing, too, that state decisions over technology should remain politically neutral choices based on technical merit, Gonzales challenged, "If Open Source software satisfies all the requirements of State bodies, why do you need a law to adopt it? Shouldn't it be the market which decides freely which products give most value?" Crucially, the account he delivered of a future with FLOSS predicted conditions of economic and technical instabilities and the devastation of what were presumed to be otherwise healthy political processes. Whereas Proposition 1609's account of an "illegally" operating government stressed the condition of generalized software piracy that resulted from adherence to proprietary technologies, Gonzales's evocation emphasized the violation of the "natural" laws of free market enterprise that governments' adoption of FLOSS solutions threatened.

Introducing the possibility of FLOSS use in Peru's government, therefore, produced an array of new globally circulating narratives around the possibilities and processes of national technology procurement. Such accounting practices not only insisted on the politicized nature of state use of either open or proprietary technologies—pronouncing how such usages could either extend or diverge from established relations of dependence—but they also rendered FLOSS visible as an alternative that could similarly either disrupt or enable global relations of inequality. Critically, the initial generation of a new interpretive account of technological possibility through FLOSS's introduction forced Microsoft to acknowledge the challenge to the established system of software procurement and to generate its own counternarrative to define FLOSS instead as a technology of risk. If prior to the visible emergence of a FLOSS proposal, proprietary closed software had appeared as the uncontestable, natural option, following the FLOSS proposition, Microsoft had to resituate itself within the language of national security, economic stability, and rational choice.

The emergence of FLOSS legislation thus proved to operate as a generative force, multiplying the narratives around code, national IT markets, and the interpretations of what it would mean for Peru's public and private sectors alike. The emergence of such accounts, however, did not simply occur spontaneously but was bound to other narrative acts engaged through FLOSS's advocacy network. To trace these narratives back was to uncover a diversity of accounts rather than a single explanatory thread that ran through them all, tying them neatly together, edge-to-edge. Among them were some, in fact, that according to their tellers had been brought into existence by complete accident.

Accounting for Local Government

Jesus Marquina Ulloa, an independent technological consultant for Peru's government, first brought the notion of FLOSS to the attention of Congressman Villanueva in 1996. Marquina, age forty when I spoke to him, recalled that his work in the mid-1990s involved routine visits to municipalities that hired him to implement tax administration software. Although he had originally developed the tax application using proprietary development tools, he had begun coding an equivalent FLOSS application that he hoped would replace the proprietary version. Villanueva, then Andahuaylas's mayor, expressed interest in learning more about Marquina's proposals. They discussed first creating FLOSS applications for municipalities or organizing an association to offer basic services to local governments and began to consider proposing legislation on a national scale when Villanueva's election to Congress in 2000 presented an opportunity to generate a new, nationally scaled action around the issue.

Still, it was Marquina's ties to local government that anchored his investment in FLOSS. He recalled that in his first years programming for regional governments, he used pirated copies of Microsoft's Visual FoxPro development tools and the database development language Clipper to develop his applications. He remembered that the primary challenges he encountered were not technical, programming problems but involved national and local governments' general treatment of data and information. From his point of view, "There is a complete lack of standards in municipalities' tax procedures. . . . Even though laws exist around taxing, they're truly very ambiguous . . . only a few officials seem to understand it—so many municipalities have their own interpretations."

Coding technological systems for local governments made Marquina increasingly aware of the broader social conditions technologies were embedded in. He described initially having to develop separate versions of his application for each of the five municipalities that contracted him at the time: "[Each system] would have to be personalized since each municipality's procedures are so distinct. . . . And I had to administer 100% of the source code by myself. . . . In the long term, the maintenance of the system becomes unsustainable." He cited with excitement his discovery of FLOSS and its principles of open, shareable, and legally modifiable code as a solution to local governments' technological dependence. He described how providing municipalities access to program source codes allowed them to locally articulate and innovate their own technological solutions: "[By] openly distributing code, instead of harming myself . . . I actually started to enjoy benefits. . . . I started to see that in some municipalities, the responsible team would administer solutions to [technical] problems that would never have occurred to me, and with a rapidity that . . . was optimal."

Explicitly highlighting his experience of confronting internal disorganization and nonstandardization within multiple municipal administrations, Marquina argued for FLOSS's use by local government offices as a technical solution to a problem he read as socially and contextually grounded.

To Marquina, FLOSS's accessible source code empowered software developers and the local offices they worked for by allowing them to legally customize code for diverse needs and evolving uses—many of which may not have been anticipated when technological systems were initially established or recognized by network actors residing far outside of local sites of deployment.

But Marquina realized he was not alone in his frustrations with the limitations of proprietary, closed software. He had joined an Argentina-based mailing list to discuss issues of FLOSS in government shortly after his own discussions with Congressman Villanueva had helped to yield Proposition 1609 in the final months of 2001. The critical relations he cultivated from what could have otherwise been described as casual activity on the mailing list generated new extensions to Peru's FLOSS advocacy network.

Summoning Citizens, Democracy, and Code

Several hundred miles outside of Lima, the Cordoba-based Via Libre Foundation, an NGO addressing concerns around technology and civil society, had already been deeply involved in promoting FLOSS legislation in Argentina. The director of Via Libre, Federico Heinz, was instrumental in building relations with Argentinean congressmen to sponsor the legislation in his home country. Also critical, he had played a central role in building discussions around FLOSS in governments through a mailing list he helped to found in 2000 called Proposición. The list would in turn provide a critical mass of concerned regional citizens whose debates contributed to the evolution of the Argentinean bill. Recalling the processes around the bill's construction, Heinz referenced the processes of FLOSS construction, where online communities of programmers openly critique and exchange pieces of code to collectively build and refine a working application. Heinz described the ten-month-long process of collective authorship: "We [at Via Libre] had hacked up some text and we brought it back to the list for it to be criticized until we reached something that was acceptable. . . . It was an amazing process really, like a participative method of creating the law—with people stating how they would like to be ruled . . . this construction model of creating legislation as if it were software."

Although originally begun with the intention of promoting FLOSS in Argentina, the list and its participants—many of whom resided outside South America—grew to address the growing phenomena of FLOSS legislation by governments worldwide, particularly among nations of the Global

South. Among the participants of Proposición who had approached Heinz about a local bill that he had helped to initiate in Peru was Jesus Marquina. Heinz, another Argentinian participant with Proposición, Enrique Chaparro, and FLOSS advocates in Argentina were recruited again in Peru's efforts a few months later to respond to the Microsoft indictment of Proposition 1609. Heinz recalled that the primary challenge was managing the flood of feedback received from participants who all attempted to deliver their independent contributions simultaneously: "This kind of thing is a lengthy process. . . . Everybody was contributing ideas and we had to continually write them down and make changes. And [then] when we posted the letter draft to Proposición, the people from the list had something to say about it [again]!"

Nearly six-thousand words long and filling twelve, single-spaced pages once completed, the response to the Microsoft letter produced from the collective efforts of the Argentinean advocates, Proposición's international participants, the Peruvian congressman and Marquina, expanded on the arguments asserted in the Peruvian bill. It meticulously enumerated and refuted each of Gonzales's assertions. It reasserted the justification for Proposition 1609, specifying that the bill was not motivated by economic rationales but by "fundamental" obligations of governments to citizens. These included ensuring citizens' free access to public information and ensuring the permanence of public data under the rationale that if state's were dependent on closed, proprietary software and were unable to afford proprietary software updates, public data would be compromised: "The State archives, handles, and transmits information which does not belong to it, but which is entrusted to it by citizens. . . . The State must take extreme measures to safeguard the integrity, confidentiality, and accessibility of this information."

In speaking with me, Heinz emphasized the unexpectedness of the letter's final form and the evolution of ideas generated through the proposition's dynamic debates: "What happened in the letter was an incremental process. [At] the start of the discussions, we looked at free software as a way to help government use software for less money. . . . But gradually . . . we discovered that free software, *even if it were* more expensive [to maintain and implement] than proprietary software, public administrations *must* use it—that it is the only way [government] can fulfill its duties." Heinz explained that a consideration of states' responsibilities to their citizens surprisingly brought the group to reconsider their own framing and arguments for FLOSS, even as well-developed as they already were: "[We realized] better software and lower cost may be necessary for a corporation, but . . .

corporations just have to be accountable to shareholders. [Citizens] are all shareholders, though, in the state—and it's not like a corporation where we can choose not to be. Cost is important but it is only secondary. When we began to think about the possible insecurities in government systems that store [citizens'] personal data, and the way this data is handled, [we realized that] *I* as a citizen have an interest in how this is guarded."

As crucial for Heinz was revealing the relationship between information technology and the politics of governance. Stressing the centrality of technological processes in shaping shared political infrastructure, he argued that limitations in technological and government transparency were directly related: "We are already waist deep in the information society—but . . . most people say all this free software stuff is just relevant to a bunch of geeks. . . . [But] software *is* a very important part of democracy. There are whole arenas, that no matter what the law says, if the software is implementing the law, software has the upper hand." Significantly, too, he stressed his own differences with what he saw as a dominant position within the larger FLOSS movement, which he characterized as focused on legal reforms for the benefit of technological evolution. He reported that he had trouble explaining the legislative strategy to a wider FLOSS community concentrated in the West and attributed such difficulties to the movement's primary concern for FLOSS market extension and reforms of Western software patent and copyright laws. Although he supports such agendas, he emphasized that he saw other crucial objectives in FLOSS. "[T]his has to do with *citizenship* and software. . . . What we are trying to achieve is not just [better] software, but a more sustainable society. . . . And using free software is a tool in building that."

Heinz's accounting of the emergence of FLOSS legislation was not propelled by the same ideals of technological evolution evoked by dominant FLOSS concerns in the West or by the user-centered localization concerns that Marquina evoked. Rather, his narrative of FLOSS's imperative was anchored in notions of citizens' democratic rights in an emergent information society. Notably, Heinz drew from FLOSS principles that promoted individual consumer freedoms and user rights and resituated these in the realm of collective political rights. To explain FLOSS, for Heinz, became a way of critiquing contemporary political and economic structures and imagining the possibility of a more democratic, participatory public. The narrative he constructed, in other words, was one built around the emergence of an information-based society in which new sites of politics and governance could come into articulation—and in which one such crucial site would be recognized as manifesting around technology and code.

Articulating Identities of Possibility

While the exchanges among Congressman Villanueva, Marquina, and Via Libre were unfolding, another body of local FLOSS advocates, the Peru Linux Users Group (PLUG) in Lima, began to dedicate resources toward the bill's promotion. Founded as an online mailing list in late 1997, PLUG serves today as a virtual community of fifteen hundred members that exchange technical information related to FLOSS use. Cesar Cruz, a coordinator for PLUG's listserv and a thirty-year-old Linux instructor, explained that he met Marquina at a Linux conference several years earlier. He recalled casually hearing of Proposition 1609 from Marquina by phone. After that conversation, Cruz decided to leverage PLUG's user base to launch an independent local campaign to promote the bill.

After debating online how PLUG would support Proposition 1609, members independently began the work of outreach, e-mailing contacts and news outlets inside and outside Peru. These publicity efforts brought news of the Peruvian bill to a broader international audience after copies of Proposition 1609, the Microsoft letter against it, and Villanueva's response were posted to the PLUG website in English and Spanish in April 2002. *Wired* magazine published their account of the Peru movement shortly after in late April 2002 (Scheeres 2002). This was followed by stories in UK-based tech news publications, *The Register* (Greene 2002) and vnunet.com (Williams 2002), and in the FLOSS news sites, *Linux Journal* (LeBlanc 2002) and Slashdot.org (2002a, 2002b).

Although international media coverage of Proposition 1609 spread within Peru, little if any mainstream news coverage appeared. Hoping to correct this, PLUG undertook various nationally targeted activities, including distributing fliers on the street corners in Lima, hanging posters on public walls and buildings, and posting to Spanish-language FLOSS news sites about Peru's legislative developments. PLUG members also organized a conference, "Linux and Free Software in the State," that featured Villanueva, Chaparro, Heinz, and the Mexican FLOSS advocate Miguel de Icaza. Crucially, all of PLUG's activities were planned independently, without any direct communication with Congressman Villanueva, Marquina, or Via Libre. PLUG members also began to invest their own savings to fund activities. As Cruz explained, "I don't know how much money I spent [to promote Proposition 1609]—but neither do I care if it was a lot. . . . Thanks to learning Linux, I've never been out of a job. . . . [Here], it's difficult to find work, and if you do, it's often with a low wage. Thanks to Linux, I've done better than the majority of young Peruvians that are dedicated to computer [work]."

Narrating his experience with FLOSS as allowing him to avoid the employment difficulties commonly found by young technicians in the national job market, Cruz minimized the expenditures he made. For him, financing PLUG's events allowed him to support what he credited for his professional security. More than merely providing the material resources to further FLOSS's general growth, Cruz also read his support as directed toward FLOSS development in Peru. Projecting a future of greater national and economic sovereignty built on the talents of local technicians, Cruz explained, "The most important thing to me is that we can develop our own technology. Before, we didn't have any possibility of this with proprietary software. Now with free software, yes, we do. . . . In my country, there are few [software] developers [because] one always buys programs from abroad. There's a very large dependency on Microsoft, and we have to break this. Because we have the capacity to construct our own software . . . more than being about nationalism, I would say that this is about being able to get out of [a situation of] under-development."

Antonio Ognio, another PLUG list coordinator and the twenty-eight-year-old founder of a Linux server company in Lima, similarly narrated his experience with FLOSS as one that expanded his professional opportunities. Recalling the last few years of his undergraduate education in systems engineering in Peru, he remembers being instructed to focus only on business administration skills, rather than higher-level operating-system programming. Feeling "extremely bored" and having an "urgent need to re-invent" himself, Ognio recalled that discovering FLOSS allowed him to pursue his own personal "goals and dreams" rather than goals that had been prescribed to him: "Why not think you can learn . . . the kind of 'secrets' hackers would? Linux put me quickly in contact with all of that."

Pointing to the limitations they see as institutionally imposed on Peruvian technicians, Cruz and Ognio build narratives around FLOSS that stress overcoming professional limitations and imagining new possibilities for growth. Discovering FLOSS for them permitted self-transformation and empowerment that distinguished them among their peers in Peru. When they imagined it at the level of national use, then, FLOSS was read as offering new potentials for developing other pathways toward technological, economic, and political independence. FLOSS developer David Sugar (2005) echoes such hopes: "In providing opportunities for Latin American citizens to directly participate in the worldwide commercial software market locally, free software offers incentives for forming a local software industry that can then compete on an equal basis with that of any other advanced country in the world."

Such interpretations of FLOSS are indeed distinct from those that Marquina stressed in his account of local government experiences or Heinz's account of citizen rights. Although the diverse accounting practices among network participants emerge independently of one another, their distinctions aren't intended to contradict each another as much as invite their mutual coextensions. Diversity in the FLOSS polyvocal network appears here as a generative force rather than a competitive one.

Repercussions: Microsoft and Other Interventions

By early summer 2002, following the wave of media coverage around the Villanueva-Microsoft exchange, several new developments in the Peruvian legislative efforts would unfold. Two new versions of Proposition 1609 were introduced to the Peruvian Congress and two additional FLOSS bills were also proposed. One would establish a consulting commission to study and authorize government use of FLOSS. The second would mandate FLOSS use by businesses whose primary client was government.

The proliferation of official FLOSS support prompted Aldo Defilippi, the director of the American Chamber of Commerce of Peru, to write a letter to the president of the Peruvian Congress, decrying the bills for "discriminating against" proprietary software companies. Defilippi's letter was followed by the US ambassador to Peru, John Hilton, and his letter, which warned that economically excluding companies such as Microsoft would hurt an industry that created thousands of local jobs. An in-person meeting between Peruvian president Alejandro Toledo and Microsoft head Bill Gates in mid-July 2002 effectively delivered on what Defilippi's and Hilton's letters had hoped to achieve. In mid-July, Peru's president met the Microsoft chairman in the company's corporate headquarters, where the leaders signed an agreement for Microsoft support of Peru's Project Huascaran, a digital education initiative providing Internet and computer access in Peru's rural schools (and the predecessor to Peru's OLPC program). It also gave Gates the opportunity to present Toledo with a donation of more than half a million dollars in software and consulting services. As the Microsoft (2002) press release for the event explained, "Microsoft will not be supporting Peru in its [digital education] Huascaran Project alone, but in other important modernization projects of the public sector and nation. . . . In addition to the support that Peru will receive for the Huascaran Project

. . . Microsoft will also design and execute an Electronic Information System with the public sector, for better internal communication and more transparency in the services offered to Peruvian citizens, and will put into motion a practical tendency for the most modern countries of the world: e-Government. "

Without a trace of the defensiveness and alarm that characterized his first letter to Villanueva, Juan Alberto Gonzales added his endorsement of the agreement in the press release, characterizing Microsoft as a responsible, corporate "civic actor" in the process: "Microsoft Peru knows its role in society, and we know that only an informed society will achieve development; and we feel that our function is to provide society with the technological resources that will permit the spreading of access to information to allow the creation of professional personnel and the development of its businesses."

Notably absent from either Microsoft's or Peru's official comments on the accord and donation, however, was any mention of FLOSS or the various bills supporting it in the Peruvian Congress. Speculation began to emerge, however, that despite all official pretenses, the Toledo and Gates meeting and the Microsoft donation were meant to secure rejection of the pending FLOSS proposals. Or as *The Register*'s John Lettice (2002) mused, "Where President Toledo's education and e-government deal [with Microsoft] leaves the Free Software initiative[s] is not clear. But as he must surely have a contract with Microsoft, it likely complicates [them]. . . . When major Microsoft contracts or customers are in peril, Bill [Gates] is frequently deployed as the last weapon."

But the visibility of Peru's FLOSS efforts invited other independent contributions with more welcome effects for its supporters as well. A 2002 study completed by the University of Maastricht's International Institute of Infonomics that was funded by the European Commission on FLOSS development and use in the public and private sectors drew its policy recommendations from the text of Proposition 1609 (Ghosh, Glott, Krieger, and Robles 2002). Such borrowings occurred without coordinating with the original authors of the text. Reflecting back with surprise on the unexpected mobility and impact Proposition 1609's content had, Federico Heinz underscored the ease by which diverse adoptions and translations occur within FLOSS networks, "It is very hard to do anything in free software that actually has any respect for national borders. Because you start doing something and other countries and places start picking it up and it becomes international in and of itself." Heinz added that as news of the legislative efforts in Peru spread globally, government officials from across Europe and Latin America began to approach the parties involved to explore similar initiatives.

Shortly after the Microsoft donation was presented, as well, officials from UNESCO approached Congressman Villanueva with plans to organize an international conference on FLOSS and Latin American governments. Held in Cusco, Peru, in 2003, the conference featured tracks on international politics and FLOSS, FLOSS's global market, and FLOSS in education, science, and culture. And through less-official channels, as well, the Peruvian legislative efforts would bear new impact. Following the media coverage of Villanueva's response to Gonzales's letter, FLOSS supporters from across the world began to contact PLUG and Pimiento, a student-based FLOSS group in Lima that donated computer servers to support the growing online demand for the documents related to the Peruvian efforts. Supporters volunteered their skills to translate Villanueva's letter into over a dozen languages, including Chinese, Turkish, Greek, Hungarian, and Portuguese, allowing the Peruvian case to acquire new mobility and audiences in each local reproduction.

Perhaps not surprisingly, recognition as a network invited a diverse array of new global interventions that reacted in multiple ways to the network's discursive productions. Investments of money, time, institutional resources, and personal skills collided with other engaged activity, hoping to redirect local network dynamics. For some investments, as with Microsoft's, it was the intention of containing new local activities and extensions, and a will to re-anchor the once-stable global relations that drove them. For others, it was to amplify the audibility and content of the network that propelled them. And for others still, it was reconducting the dissonant chorus of voices in the network that motivated them. Yet that such attempts to diminish, amplify, or newly channel the network saw themselves as necessary bespeaks the collective force of the discrete voices flowing through it. However internally disorganized, undefined, and cacophonous Peru's polyvocal FLOSS network may have been, it was still a body that actors found they could not afford not to react to.

Speaking for and through the Network

Science studies brought early attention to the notion of networks as formations of social association as well as bodies of political representation. Bruno Latour and Michel Callon stressed how networks emerge from the work of successfully localizing or "translating" technoscience from centers of production outward toward distributed, distinct contexts. Such translations would then call on a particular representative actor to unify interests through functioning as a spokesperson for the whole (Callon 1986; Callon and Latour 1981; Latour 1987, 2004a). Such scholarship specified that in order to allow a single actor to emerge with the authority to speak or act on behalf of other actors, the work of local translation must be followed by other acts of "persuasion and violence" that would be amplified through the unification of diverse positions: "Whenever an actor speaks of 'us,' s/he is translating other actors into a single will, of which s/he becomes the spirit and the relationship between spokesman. S/he begins to act for several, no longer for one alone. S/he becomes stronger. S/he grows" (Callon and Latour 1981, 297). Networks here channeled and manifested authority for the actors that are interconnected through them, allowing a single actor to speak and act for a multitude distributed across local contexts and to "lay down a temporality and a space that is imposed on others" (Callon and Latour, 1981, 287). So that if "before, the elements dominated by the actor could escape in any direction . . . now this is no longer possible. *Instead of swarms of possibilities*, we find lines of force, obligatory passing points, directions and deductions" [emphasis added] (1981, 287).

Science, technology, and society (STS) scholars also pointed to the ability of scientists to "speak for" and politically represent the diverse human and nonhuman actors linked together through networks (Callon 1986; Callon and Latour, 1981; Latour 2004a). Their role as spokespeople for the natural world grants them a unique authority, "endow[ing them] with the most fabulous political capacity ever invented: They can make the mute world speak, tell the truth without being challenged, put an end to the interminable arguments through an incontestable form of authority that would stem from things themselves" (Latour 2004a, 14). In their unique role as spokespeople for the mute, natural world, scientific experts, despite their limited numbers, are able to persuade wider circles of external actors and recruit new allies and resources. The greater the number of allies recruited, the further the network extends, and the greater the strength and stability the network gains. Through such strategies of diverse recruitment and representative unification, Callon and Latour (1981) specify, "a handful of well-positioned men of science may rout billions of others" (181).

Such a unifying function of spokesmen, however, was notably absent in Peru's polyvocal FLOSS network. Although network spokespeople operated, following diverse technoscientific translations, to unify local accounts of "truth" for actors across distributed networked spaces, the discursive practices of Peru's FLOSS advocates functioned to challenge what were considered established facts and shared presumptions around software espoused by dominant network actors. In the process, they multiplied explanatory accounts. Network spokespeople might be employed to minimize "swarms of possibility" in favor of constructing "obligatory passage points," but it was precisely those new possibilities that Peru's FLOSS advocates gave voice to. And although network spokespeople, as actors authorized to represent diverse voices, should allow a few centrally positioned actors "to dominate from a distance" (Latour 1987,243) and control peripheries from centers, Peru's network actors disclaim any authority to speak on behalf of the whole and instead continually reference the contributions of one another in acts that distribute credit.

Susan Leigh Star and James Griesemer (1999) likewise characterized scientific networks as constituted by a heterogeneous ecology of institutions. They stressed, however, how much of the work of scientific networks is performed by actors other than the scientists at the center of the network. Emphasizing the contributions of participants situated outside the realm of professional science, the ecological approach they argued for "does not presuppose an epistemological primacy for any one viewpoint. . . . The important questions concern the flow of objects and concepts through the network of participating allies and social worlds" (507). Such an approach, the authors asserted, made visible a "many-to-many mapping, where several obligatory points of passage are negotiated by several kinds of allies" (507), who are tied together not by consensus-seeking immutable mobiles but by flexibly interpreted boundary objects.

Highlighting the capacity of boundary objects to coordinate the activity of actors across distinct sites, the authors explained them as objects of technoscientific engagement shared across several intersecting worlds, "satisfying the informational requirements of each. Boundary objects are both plastic enough to adapt to local needs and constraints of the several parties deploying them, yet robust enough to maintain a common identity across sites. . . . They have different meanings in different social worlds but their structure is common enough to more than one world to make them recognizable means of translation" (509). There was no need for a spokesperson to represent the united voices of the parties who are organized through boundary objects. Actors instead maintained the integrity of their own interests, whereas networks around boundary objects accommodated the diverse interests of those actors linked through them. More than representing the consent of the diverse parties interlinked together as a unified and ordered whole, networks here expressed the pragmatic will of such parties to coordinate their activities in the interest of advancing a particular, shared goal. Diverse social worlds may organize themselves around boundary objects but their coherence was not one that necessarily predicts long-term stabilization or the durability of a particular vision of reality. The network's coherence rather represented the intersection of outside social worlds and expressed the will to maintain social unity for sustained or short periods of time according to the needs of particular actors. And although relations here were not dependent on a spokesperson, it was the capacity to unite the interests of diverse actors around a shared, central object and to generate coherence that explains the network's strength. How *extension* or the outward, dynamic growth of networks developed, however, was less visible under approaches to boundary objects than their ability to draw in, stabilize, and center diverse, distributed interests.

Such accounts of network center–periphery relations provide valuable insight into Peruvian FLOSS advocacy networks for highlighting how technological innovation emerges not as the result of individual genius—of a lone congressman, for instance—but through the participation of multiple interlinked parties and negotiation of diverse interests. These parties in the case of Peru's FLOSS advocacy network included not only Peruvian congressmen but also civil society–based NGOs, independent coders and programmers, and government-employed technicians. Assuring that such groups continue to perform simultaneously as a network is key to the sustained advancement and audibility of FLOSS advocacy. Still, it is the key work of enabling central anchors and the value of constructing a common shared project among multiple actors that is stressed through boundary object accounts. The dynamics of outward flows, extensions, and regenerative work from peripheries themselves—including the value captured in the range of differences expressed by network actors—such as those participating in Peru's polyvocal FLOSS network, however—remains here less visible.

More recent research on networked organizations build on these theoretical foundations of networks as representative bodies, stressing how the spread of new information technologies allows diversely situated, geographically dispersed actors to act collectively in distributed social formations. Research on the proliferation of FLOSS communities in particular has been influenced by the notion of networks as collective bodies that emerge from the varied individual interests of its participants (Benkler 2006; Coleman and Hill 2004; Kelty 2005b; Lin 2007; Weber, 2004). Chris Kelty (2005b, 2008) describes networks of geek socialities as forms of representation that produce a language, folklore, and technical code. Describing geeks as constituting a global and "recursive public," he writes that the diverse actors associated with geek communities—including hackers, lawyers, activists, and IT entrepreneurs—are networked together by their shared concern for the legal and technical possibilities for their own association. Yuwei Lin

(2007) similarly points to the productivity of diversity in FLOSS communities, writing that heterogeneity serves as "the resource that helps mobilize the FLOSS innovation [and that] drives diverse actors to re/define and practice the hacker culture they perceive differently." Coleman and Hill (2004) underscore the multiple social groups and interests that organize themselves around FLOSS technologies. That such diverse parties as large, transnational technology corporations, anticorporate political activists, and technology hackers share a common interest in FLOSS demonstrates its ability to perform as an "iconic tactic," a strategic practice that is productive of other social and political practices. For legal scholar Yochai Benkler (2006), such diverse tactical uses of FLOSS are to be expected given that the networked information economy is built on the enhanced autonomy of individuals who are now given a "significantly greater role in authoring their own lives" (9). He, similar to political scientist Steven Weber, points to the surprising economic and technological productivity that results from the aggregate of independent, voluntary acts of dispersed individuals (Benkler 2006; Weber 2004).

These more recent approaches frame networks as social formations that accommodate and foster political diversity and cultivate new means of political action and expression. Rather than stabilizing reality, these networks—much like Peru's polyvocal FLOSS network—generate a plurality of discursive spaces and practices that call attention to new, still unrealized possibilities in law, economics, and culture. Networks here are not only bodies that represent and "speak for" the collective of social and technical actors intertwined within them but also seek to express a collective will pressing for distinct futures, challenging the conditions that structure contemporary life, and seeking to effect shared social benefits and political change.

Such attention to the productivity of networks in generating new cultural codes and discursive practices is echoed in the work of social movement theorist Alberto Melucci, who emphasized the difficulty in identifying a single operational logic to networks. Using the diffuse, fragmented structure of contemporary forms of collective action as his network model, Melucci (1996) writes, "One notes the segmented, reticular, and multi-faceted structure of movements. This is a hidden, or more correctly, latent structure; individual cells operate on their own entirely independent of the rest of the movement, although they maintain links to it through the circulation of information. . . . Solidarity is cultural in character and is located in the terrain of *symbolic production* of everyday life" (115). Constituted by a composite of diverse and even contradictory elements, networks here resist collapsing the plurality of their actors under a single uniformity. Absent is a center of control or single explanatory axis. Emerging instead are diverse, individually constituted goals that integrate themselves into and reinforce the network.

Networks here don't explain so much how social coordination or technological standardization is achieved as much as how meaning emerges and multiplies from the noise and cacophony of an ambiguous "symbolic field." Networks of collective actors operate not to distill or filter social ambiguity and complexity but to make such elements evident and to unveil the taken-for-grantedness of naturalized categories. That social processes appear as standard and ordinary at all is an effect of taken-for-granted political work that actors seek to reveal. "Bearing the banner on spontaneity, purity and immediacy of natural needs," Melucci (1996) asserts, "contemporary movements move to challenge the social and its reduction of differences to systemic normality" (96). Through their discursive generativity, network actors pronounce the artificiality of the natural, and make possible the recognition of the "abnormal" as new potentials.

Such attentiveness to networks as discursively productive—and marked by their ability to create new cultural meanings to contest power and produce alternatives to dominant codes—has similarly been expressed by Latin American cultural studies scholars. They stress that by advancing alternative concepts of civic identity, citizenship, and democracy, social movements succeed in unsettling dominant cultural meanings and political narratives and ultimately create new public spaces for collective protest (Alvarez, Dagnino, and Escobar 1998; Canclini 2001; Eckstein and Merino 2001; Fox and Starn 1997; Yudice 1998). Such authors highlight how network practices react to dominant constructions and create new cultural extensions in the work of destabilizing what was established as "common sense" through their new discursive productions.

Network studies have approached networks as bodies of representation that unify diverse actors' voices and attend to the plurality of voices that emerge from and are fostered by such formations. Networks may serve as bodies that unite the social interests of diverse groups but dependency on spokespeople is far from given. In part, this can be explained by the fact that it is cultural change and not primarily technological stabilization that serve as organizing motivations of the polyvocal network. Likewise, it is not the generation of a dominant, universalizable notion of truth and knowledge that network activities are channeled toward. Rather, participants' activities animate and extend the kinds of "situated knowledges" feminist science studies scholar Donna Haraway (1991) highlights. Such multiple microknowledges, which frequently belong to those who have been denied political representation and privilege, insist on a form of "seeing from below" (192) that draw attention to the partiality of universalized notions of truth and in doing so destabilize dominant conceptions of knowledge. They also insist on a politics of the local that may be connected to other networked and global trends but never forget the material "situatedness" of the bodies and actors participating in global networked forms.

Indeed, Peru's polyvocal network underscores that central to network performance is not merely the production and extension of the technological artifacts of FLOSS but also the production of new cultural meanings, discourses, and narratives that account for the technical *and* political value of FLOSS within local contexts. Among the varied practices that Peru's advocates construct around FLOSS, it is the absence of a desire for a uniformed, universal standardization or a purposefully managed coherence that one notes. Their voices aim to highlight the limitations of established social and technical practice and stress the potential to pursue new and distinct possibilities. Not surprisingly, the visibility of new alternatives likewise retains the potential to affect new, frequently unanticipated repercussions. Perhaps what the unpredictability of these network effects makes clear, then, is the utility of attending closely to the individual voices interlinked by networks and of recognizing how network strength may be measured not merely in the degree to which uniform expression is achieved. Rather, strength might be assessed too in the degree to which such uniformity is undone, and independence from the network spokesperson proclaimed.

There was little about the events that surrounded Proposition 1609 and its eventual passage as a technological neutrality law that its Peruvian advocates considered inevitable. In asking Peru's advocates to reflect on the outcomes of their efforts, it is a distinct lack of consensus that one hears about what was achieved. Some, considering the long wait for the law's passage, the rejection of the other proposed FLOSS bills, and the altered language of the original bill from obligating government use of FLOSS to obligating a stance of neutrality, lament that efforts among distinct parties could have been better coordinated. Antonio Ognio, for instance, told me that "we showed a great commitment and always supported with very specific goals . . . but were lacking the time and conviction to sit down and talk about strategies and mid-term plans. . . . We got the 'geek' community involved in politics . . . but we've failed to have it organized and go."

For others, however, it was the ability to have built and generated international publicity around the Peruvian efforts at all that proved crucial.

Emphasizing the contribution made in simply having revealed new possibilities in social and technological conditions in Peru, and for bringing this to the attention of international audiences, Cesar Cruz told me, "Each effort we made . . . was important. . . . because even if the bill hadn't passed, we've managed to make people pay attention to us and Linux. . . . The principal thing in this moment is that each time more people know that there exist options, because before, everyone believed that Microsoft is the only technology, which just isn't true."

Filtering distinct events, effects, and intentions, the incongruencies between network actors' modes of reassessment demonstrate the resistance to a single explanatory account imposed on the network. In the case of Peru's FLOSS legislation, the network operated not so much in the interest of standardizing expression, meaning, and practice but to generate a multiplicity of accounts around FLOSS that expressed a heterogeneity of forms. Some stories, such as Villanueva's letter, traveled widely and freely beyond the scope of their producers, adopted by other audiences and serving as a catalyst for new modes of practice and analysis. Crucially, as well, the production and dissemination of such narratives became revealed in collective, distributed acts in which multiple parties independently participated in and contributed to effects that they note they were only partly to credit for and only partly controlled.

That these isolated explanations of activity on the network might appear incomplete and unfinished, necessarily referencing other autonomous streams of activity in the network to explain themselves, however, is a sign that theoretical accounts of network productivities may themselves need to be diversified. Peru's polyvocal FLOSS network urges us to consider the disparate meanings and discursive practices that flow around FLOSS as central to a network's performance. It prompts us to consider how it is not merely the production and extension of new technological artifacts such as FLOSS that the strength of a network depends on but also the production of new cultural meanings and narratives. It reveals, in other words, how networks may function powerfully not just through their technological innovativeness and scientific productivity but also through a cultural and discursive extension that, however disordered, noncoordinated, and cacophonous, may nonetheless produce wider political effects. And it demonstrates how such productivity may be generated, even after having retired the network's spokesperson.

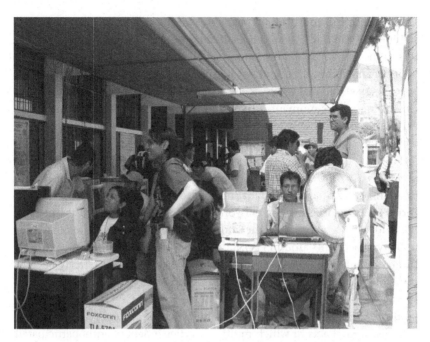

An outdoor installation festival organized by the Free Software Student Association at the National Engineering University (La Universidad Nacional de Ingeneria) campus in Lima in 2006. The event, which drew free and open source software advocates and users in from across the city, also featured a two-day long series of talks and the free distribution of compact disks of free software applications.

5   Recoding Identity: Free Software and the Local Ethics of Play

Antonio Ognio, a twenty-eight-year-old Peruvian programmer living in Lima, recalls his early days as a free software novice in the late 1990s. He had become vaguely familiar with a growing alternative to Microsoft's Windows operating system, Linux, from taking software engineering classes at a private university in the capital city. But at that point, as he puts it, "The only thing [my university classes on Linux] accomplished was getting me and my friends to crash a PC by coding very bad C programs while learning fork commands." His curiosity about the then-relatively obscure FLOSS phenomena wasn't really piqued until he opened a copy of the US-published *Wired* magazine and saw some screen shots, featured across its glossy pages, of one of the popular FLOSS desktop environments, Gnome. "I was amazed. I knew nothing about free software then. I was aware of the existence of Linux but had no clue it was free, especially [free] as in [freedom of] speech. I just remember feeling I *had* to have that cool desktop running in my home computer ASAP. The art was cool, but I'd say [it was really] the challenge of going into this whole new hackerish world that was so exciting." He remembers immediately boarding a bus headed for downtown Lima where he purchased a disk of free software for three dollars that he installed on a home computer. And although he relays this account to me through no more than the compact digital window of an ICR channel, the enthusiasm he remembers feeling back then is still clearly conveyed: "It kept me up the first night and the next day and then the next day learning about partitions, boot managers, devices, packages, shells, tarballs, sources, and everything!"

He leaves little doubt of the impact that episode would have on him. When I meet Ognio in person a few years following, he's become well recognized by family and friends inside and outside of Peru for his avid FLOSS use, advocacy, and entrepreneurship. He had switched his computer's software from Microsoft's Windows to free software applications years ago and had begun making unpaid code contributions to a number of well-known global projects, including Apache, the popular free software server application, and Gnome. Even non-techy friends knew of the work he had done to help found a nationally anchored group of Lima-based free software users who had worked to sponsor free software legislation and press for the mandatory adoption of free software in public offices. Working with organizations such as PLUG[1] and APESOL,[2] activists had collaborated with politicians to introduce one of the first pieces of national legislation for government use of FLOSS. Similar legislative movements had begun in a handful of other Latin American nations at the time, including Brazil and Argentina but it was the efforts in Peru that unexpectedly garnered international publicity after a donation from the Microsoft Corporation in the summer of 2002 generated controversy just months after Peru's Congress began to consider the bill.

The experience of being a young programmer with fairly typical professional goals, whose idle, early experimentations with FLOSS unexpectedly prompted a transition to becoming such an enthusiastic user and coder of FLOSS that they would revoke their own prior uses of closed proprietary software (especially Microsoft's) was an experience that multiple Peruvian FLOSS users I spoke to shared. Many would in fact become such pitched critics of standard uses of closed, proprietary software (beyond their own), and its dominant presence in and support by social institutions, that they would even be compelled to volunteer unpaid labor to advance FLOSS code projects. And many would find themselves dedicating similar degrees of impassioned investments to advocate for broader policy and institutional reforms to promote FLOSS technologies. Such radical transformations involved a distinct reorientation of self that depended on their learning to "see software"—not simply in its expanding global relevancies to internationally dispersed technology enthusiasts and cosmopolitan audiences generally—but also in its inextricable relations to local debates and concerns around national practices of information exchange.

Such processes were by no means automatic. As globalization scholar Anna Tsing underscored in studying Indonesia's contemporary environmental movement, learning to "see" the objects of advocates' commitments entailed precisely their coming to read such objects differently from dominant frameworks. As she writes of the way urban, middle-class student activists mobilized around environmentalism and reframed "nature" in the process, "The nature they love is not the fields and forests of ordinary, parochial, rural lives. As they learn to love nature, student nature lovers break away from the world of routine and authority to embrace the breadth and freedom of the outdoors. The outdoors is made modern, technical, and scientific; it must be taught in classes and taken into one's own practice through discipline and experience. . . . [It] results from a training of internal agency, desire, and identity; it is a matter of crafting selves" (Tsing 2004, 122). Much as such actors first had to distinctly "see nature" before acting on behalf of new commitments, so, too, did Peruvian FLOSS users first have to come to see distinctly through their software-based engagements before acting on such new consciousness.

Of course, that Peruvian FLOSS activists can and do use an object like software to "craft selves," speak for nationally grounded concerns, and raise local debates is significant. Software, after all, is a distinct kind of globally translating entity different from nature. And whatever similarities they might share for each having cultivated a means to mobilize contemporary audiences and charge cosmopolitan imaginaries, the difference between what it means to build a situated and critical consciousness through software, and what it means to engage in the same through an object like nature are not minor. Engagements with the natural world may still conjure notions of connecting to raw, material bodies anchored in sublime spaces of origin. But software engagements, particularly in the post-Internet age, defy such spatial and place-based imaginaries. Channeling connections to networked systems of information and rationally ordered data, mediated from or to multiple global elsewheres, software connections seem to resist the same particularities of place. And although nature might be idealized in its raw, untampered encounter, projected often as modernity's other and enabling connections to forces that preceded those of modern man, software engagements lend themselves less easily to parallel critiques of the modern. Built as they are on the manufactured products and technological triumphs of modern innovation, software is too commonly framed and uncritically celebrated as the circuit from which new global futures will necessarily extend.

Wresting software and digital products from such dominant global frameworks and drawing attention to them as not merely code-based products of modernity's triumph but also a response to its many lapses has been indeed among the crucial contributions of FLOSS communities from the Global South. Contemporary workspaces and new IT start-ups—especially in regions such as Silicon Valley and their Western counterparts—after all have been heralded for remodeling the traditionally hierarchical firm to suit the ethics and logics of production in the networked age. The youthful cultures of new media and IT companies in the United States from Google and Facebook to Razorfish and Zappos have gained renown and drawn pointed attention for cultivating work spaces outfitted more as playrooms than "offices" (Auletta 2010; Ross 2004, 2010; Vaidhyanathan 2011). Frequently showcased in popular accounts are company campuses that explicitly mix "play" with work—with fixtures such as video game parlors, movie theaters, campus bikes, gyms, and spas amply scattered across grounds.

Organizational sociologists note that such architectures embody the key innovations that distinguished contemporary "networked" organizations from traditional firms. Introducing norms and mechanisms to flatten hierarchical relations between management and creative work and engineering teams, such networked organizations were said to leverage the self-driven, independent, "self-programming," and solution-seeking enthusiasms of information and design cultures (Castells 1996, 2003; Stark 2001, 2009). By disrupting traditional workplace norms—and now drawing from emergent "creative classes" to optimize rather than hamper individual creativity and freedom in the workplace—networked organizations could appear to at last resolve the problem of uninspired and undesirable (even when not underpaid) labor (Florida 2003; Howkins 2002; Whyte 2002). Such workplaces likewise prompted digital cultural commentators to ponder if information work indeed foretold a future in which the line between work and play increasingly blurred, and in which the demise of modern capitalism's ills could at last be secured under new, digitally enlightened corporate cultures (Dibbell 2007; Himanen 2001; Jarvis 2009; McGonigal 2011; Terranova 2004).

Dependent on the individual agency of workers, flattened management hierarchies, and the active accommodation of creativity, twenty-first-century digital firms, the notion went, represented an evolved economy, one that more than ever ran on the lively, playful energies of individuals whose "work"—requiring minimal authority to manage—channeled the self-generated creative drive of contemporary hacker-inflected cultures (Himanen 2001; Terranova 2004). Yet such celebratory accounts of evolved digital labor notably base their notion of free, non-work-like work environments on a presumed contrast between the categories of work and play, where "work" represents the stultifying, managed condition of labor and "play" represents, supposedly, its stark opposite as a realm of unbridled, individually directed freedom.

Peruvian FLOSS advocates share similar hacker-inflected enthusiasms. But while they may seem, at first, to simply channel the celebratory gusto of dominant framings of digital labor, where the self-optimized enterprise of the technologically savvy are to credit for securing futures of heightened creative production and productivity, their narratives do anything but take for granted the proclaimed evidence of modernity's triumph or seek to highlight their own individual creativity alone. Their accounts instead destabilize the powerful, globally circulating frameworks of digital work as collapsing labor and leisure by underscoring distinct readings of "play" that resist its popular interpretation and celebration as the absence of "work." Insisting on other aspirations and imaginaries that channel through locally cultivated hacker ethics and socialities, they echo psychoanalytic and anthropological readings that highlight crucial work—namely, in self-interrogation and cultural critique—undertaken through acts of "play." Peruvian FLOSS advocates' identity practices thus highlight their coded creative productions and "work" less as expressions of personal ingenuity or professional resources to be leveraged as "creative classes" with networked organizations than as critical responses to restrictive institutional policies or necessary solutions to misguided ethics of local leadership. And although their narrations reveal critiques of present institutional authorities, they operate less to press for their abolition—or to celebrate the possibilities of unfettered individual choice—than to highlight the need for more accountable leadership, policies, and ethics that could foster conditions of generalized, social (and especially, education-based) benefits.

Such a notion of "play" as never having been entirely about simple fun is one digital cultural theorists have explored before. Technology studies scholar Sherry Turkle's analysis of play in digital spaces drew attention, for instance, to the "serious work" and transformative potentials embedded within play.[3] Her exploration of identity practices in cyberspace emphasized the power of play to compel change in offline worlds (1995, 2004). Through activities like building virtual worlds, interacting in online communities and engaging in multiplayer environments were possibilities of reflecting critically on the self and the real. As Turkle (2004) writes, "Cyberspace opens up the possibility for identity play, but it is very serious play. People who cultivate an awareness of what stands behind their screen personae are the ones most likely to succeed in using virtual experience for personal and social transformation" (22). Cyberplay could thus prompt a crucial process of inquiry and self-exploration around the very nature of agency and self-directed presents and futures, including, "What does my behavior in cyberspace tell me about what I want, who I am, what I may not be getting in the rest of my life?" Yet if play's potentiality lays in its ability to provoke such explorations, it could be a highly contingent one by Turkle's account. There is no external force that guarantees either that such processes of inquiry would commence or that answers would come forth in a way that could bring the self back into a stable, secure alliance with the real.

Such unpredictability and volatile potential underlie a capacity to extend individual experiences of exploration into the search for a collective experience of resistance and political opposition. Within play's space of transition—or spaces of liminality in anthropologist Victor Turner's (1967) formulation—considerable degrees of invisible social and individual work are simultaneously done. The self must come to recognize the external world as that which is distinct from personal reality, he or she must come to accept such differences and ultimately undertake the project of self-adjustment in order to reconcile such spaces and refuse other possibilities. Play, reconsidered under psychoanalytic terms as a process of interrogation (rather than activity whose ends are based primarily on fun), therefore, is revealed as a serious engagement indeed—one whose stake includes the transformation of self, the reaffirmation or refusal of the social and its ethics, and the interrogation of worlds within and beyond.

That Peruvian FLOSS programmers' initial engagements with free software as an object of playful exploration could lead to collective forms of local political action dislodges talk of digital cultural "play" from the celebratory projections of cosmopolitan "creative classes" that neoliberal framings advance. Highlighted by their narratives instead are emphases on play as engaged inquiries that press for a careful consideration of the layered obstacles that minimize or inhibit potentials for information-based critique, activity, and learning in local contexts. Their own cases of personal transformation and achievement as members of emergent information and "creative classes" who rejected standard professional goals and prescribed norms for skilled IT classes of their generation to seek out distinct applications and new possibilities with FLOSS technologies, resist neoliberal logics that often leverage such "ideal" cases of individual exception and enterprise (Ong 2006). Their self-framings emphasize instead the systematic unevennesses and routine lapses of national securities and institutional policies that surround the education and professional development of regional "information classes." Via their accounts of the challenges and tensions they faced in engaging in local FLOSS "play," they challenge notions of digital labor and the "creative classes" as the culmination of individual freedom and future-oriented enterprise, and insist on the need to reshape social worlds and ethics to accommodate distinct futures instead. Such processes of information-engaged play experienced through individual and technological exploration press toward practices of social interrogation and critique that—as Peruvian FLOSS activists' personal narratives testify to—can provide a critical means of envisioning alternative local futures.

Play and Potential

Ognio today owns his own Lima-based business that runs and offers services with Linux-based servers. It is a business he spends many daily hours and nearly every day of his week attending to in one aspect or another. In the months following the release of Peru's FLOSS bill, however, much of his free time had been spent helping to generate publicity on a regional and international scale about Proposition 1609. Serving as one of the central coordinators for PLUG, he was regularly interviewed on domestic news radio shows and by multiple technology-centered news outlets in the United States about the Peruvian bill, he helped to organize and spoke at multiple regional and international conferences dedicated to FLOSS in Latin America, and his activities at times meant entire afternoons spent passing out pro-FLOSS fliers and posters on the streets of Lima. Routinely interweaving logics of a global IT entrepreneur and local information activist into the span of a single day—as his audiences shifted among international clients, regional government representatives, and nationally based FLOSS users—he has been, undoubtedly, emblematic of the vibrant productivity of Peru's emergent information classes. And, notably, despite the hours of unpaid work his varied efforts for the free software bill demanded and his acute consciousness of the risky unpredictability of national employment markets—it was work he still gladly volunteered his labor freely for.

When I first spoke to him, it was indeed to explore the various activities that had generated international visibility for the legislation in Peru and to probe the question of what it was that drove him to dedicate so much of his energy, time, and even personal savings toward FLOSS. We were not more than fifteen minutes into our first conversation together, however, when the conversation took a distinctly introspective turn, and Ognio began to explain free software's pointed impact on his personal life and self-image. He told me, "[When] I was facing an urgent need to reinvent myself, [free software] helped me do it. . . . It was absolutely key to reinventing myself into some sort of free software, Linux-aware network entrepreneur." When I asked him to explain how he saw free software as helping him to "reinvent" himself, he responded by saying, "by letting me materialize my goals and dreams. That simply."

To elaborate, he drew me back into his past, only a few years prior, when he was a twenty-two-year-old university student in software engineering. He described his former self back then as technically ambitious but barely able to sustain interest in his classes. Feeling underchallenged and uninspired by university work, the prospect of realizing the kind of future with computers that he and his classmates were encouraged and expected to pursue was something that seemed an ever more remote possibility to him. As he put it, "I was getting extremely bored with university. I had excelled in a C programming [language] course, had also wonderful grades in a SQL[4] course, and was considered a good programmer. . . . [But the university] credentials were useless to me. I started to realize there was so much more to learn about computers . . . [than] just RAD[5] toying [that we learned in classes]. I tried to get interested. . . . It was a very strange feeling. On one hand I was amazed with the kind of perception you could get if you were reading [information-technology publications like] *Object Magazine*, and on the other hand . . . the courses made me want to quit."

Part of his frustration with his university classes was his sense of being strictly prohibited from undertaking more challenging coding and programming projects. He recalls being explicitly discouraged from pursuing more advanced programming skills by university leadership from professors to deans, and being urged to orient himself to the demands of established local businesses instead: "I was told at university that [operating system] level programming was not for us [Peruvian students]. We [were told we] should instead . . . learn enough business administration and transition to a less technical job. I clearly remember my dean telling us 'not to dream' of working with Bill [Gates] in Redmond but to focus on solving enterprise problems [for local markets]. [But] *of course* I dreamed of becoming a great programmer, and to learn as much as I could about software. . . . Why not think you can learn . . . the kind of "secrets" hackers would?"

Such an ambivalent account from Ognio—underscoring a critique of higher educational institutions' conflicted role in training for regional job markets—has been echoed by Latin American social scientists before (de Wit, Jaramillo, Gacel-Avila, and Knight 2005). Since the 1990s, as growing demands for higher education were met with policies to enable the growth of private universities and for-profit educational ventures, the public role of universities began to give way to distinctly "corporatiz[ed] university structures and cultures" (Vessuri 2011, 25). In Peru, such shifts were marked with the 1997 Act of Promotion of Investment in Education that amended the University Act of 1983, extending the right for higher education to be offered under for-profit bases and abolishing prior requirements to verify sufficient demand for creating institutions with qualified staff and resources. The new legal framework redefined private universities instead as corporations with private rights that needed to comply only with the "general specifications of the study plans" and thereby effected the rapid expansion of higher education as a consumer-centered service. Between 1980 and 2006, the number of universities more than doubled from thirty-five to seventy-eight, with the vast expansion being among private ventures. As a 2005 World Bank study noted, "These new institutions generated intense competition in the education market, resulting in a diversification of programs and degrees, but this was not necessarily an indication of meeting new requirements for professional education or knowledge production. . . . Diversification was more a result of marketing strategies to recruit students than consideration of the needs of society and the job market" (Castillo Butters, Bernuy Quiroga, and Lastes Dammert 2005). Such effects have been especially noted in the growth of IT-related and technical training in Peru, which, despite having broad presence across campuses, has still been lamented for lacking "coherence" and "regulation" in curricula design and generating "degree programs [with a] focus on the application of computing technology instead of the principles that enable its creation" (Alvarez, Baiocchi, and Pow Sang 2008, 35).

Ognio's recollections of feeling restricted, confined, and disappointed by a university system that discouraged students from developing skills to their full potential were far from singular—even if his own reaction was particular. He remembered making the decision to leave his university program in Lima and move back home with his family in the coastal city of Ica, some 300 kilometers south of the capital city. Jobless, out of school, and feeling undirected, he recalled arguing with his parents over his professional prospects and if his new and seemingly strange interest in FLOSS could possibly lead to anything more serious. He stressed that part of the disagreement involved the hope that he would follow his father's path in medicine. Despite such pragmatic pressures, he recalled his growing consciousness of a love for working with technology. He spent his next year and a half in Ica miles away from the same technologically oriented student community he had in university but still investing his time into tinkering and experimentation with free software systems and code. "That was basically when I started to teach myself Linux and PHP[6] [language] programming, [a free software web-page programming language]. . . . I got amazed by a language that would let me code ASP[7] pages but with a C-like language, so I wanted to pick [them] up immediately and focused on building database-driven dynamic websites with FLOSS technology." He explained he spent much of his day in local Internet cafés, or cabinas, whose spare, utilitarian-oriented interiors provide the primary means for local users to gain online access (typically at a cost of less than $1 US per hour). And he recalled how he would carry a pack of forty floppy disks with him "just about everywhere" that he used to store programs and ongoing projects he was working on. Saved on a few disks was code that he used to boot up a Linux partition on cabina computers, which largely ran Microsoft Windows platforms. Ognio continued experimenting with Linux-based programming, finding a global online community to help him with tips to bolster his coding skills and occasionally using such skills to contract for small jobs.

It was during that period, and despite his condition of unemployment, that he began to volunteer his own programming work, too, toward a free software-based project, contributing code to an application designed for desktop windows management known as Enlightenment. As he explained, "I wanted to *learn* the mechanics of being a free software developer, so I dedicated myself to helping to develop the Enlightenment window manager. I was a big Enlightenment fan at the time and The Rasterman and Geoff 'Mandrake' Harrison (two leading free software programmers) were my idols then. . . . [So] in the floppies I carried source tarballs, RPMs,[8] and lots of Linux [code] for [Enlightenment desktop] 'themes.' . . . [I spent my days] learning about how the mailing lists works, what the 'patches' are, about code versioning and CVS[9] repositories, about feature request, bug splashing, and API [Application Program Interface] design."

A number of his volunteer contributions can still be accessed from the applications. One early design for an Enlightenment theme that he had hacked from online archives of freshmeat.com, a repository of code for a range of free software projects, was developed to mimic the desktop environment of NextStep, a Unix-based operating system founded by Apple Computer's Steve Jobs.[10] Despite the uncertainty of the period and the lack of stable employment, Ognio still recalls the period when he first began learning, exploring, and coding for free software projects in almost nostalgic tones, emphasizing how it gave him a sense of making valuable contributions, even if he was unpaid for them, and lamenting changes in former projects he contributed to: "[Enlightenment] was a very famous free software project in those days. Now it's almost gone. It was all about making Linux and X-Windows look cool. I loved using all that spare time I had then mastering the art of programming [for it]. And part of it, too, of course, was that I wanted to 'give back' [to the project]."

He recalled, however, the difficulty he had attempting to explain his FLOSS contributions to his parents, whose practical encouragement was for him to find real paid work and not merely toy with computers: "My parents wanted me do something more 'productive.' . . . They just weren't going to believe the tales I started to tell about [one day] coming back to Lima and selling Linux services to enterprises and building dynamic websites for media customers. . . . I had a wonderful chance to learn about free software and was learning a lot that year . . . but my parents [weren't excited about] . . . let[ting] me spen[d] hours and hours doing something they couldn't understand." He remembered one incident where he even appeared to have broken a home computer that would output only pages of compiled code on the screen: "I shouldn't have left the computer [just] compiling the latest sources of Enlightenment from CVS. Since the [source code] compiler's[11] output is pretty ugly, a screen full of hieroglyphics got my dad very worried."

He explains, however, that part of his growing conviction around free software had to do with technical processes and tasks that once seemed impossible but were now practically and easily achievable: "[Discovering free software] was like finding the hidden treasure. I had spent months wondering where to find the source code to compression algorithms, network protocols, and related stuff. I was so blind that I thought that they only existed as the IP of software corporations. Having the source code in my hands and reading [free software's] General Public License[12] was like a crystal ball to me. I knew it could only be a matter of time before I'd make something of it."

He adds that free software further seemed to lift the restrictions that university leadership had erected for their programming students: "Linux was *exactly* the opposite of what my dean said was impossible: having the core of the operating system, helping build the platform that millions of people will be using. . . . Linux put me quickly in contact with all of that stuff: programming, networking, security. I switched my role model from [Microsoft's] Billborg to [Enlightenment's] Rasterman in weeks." It also allowed him to prototype a future enterprise centered on offering free software services—including building servers, web pages, networks, and database systems for Linux—before actually launching a company: "While I was picking up Linux I was also learning about all the necessary 'parts' for a business. . . . The confidence of my parents started very slowly to come back when they started seeing [I could afford to] take a bus back to Lima [for short-term] IT services." Using his earnings for contract work, he tells me he was eventually able to finance the legal work necessary to found his own company.

Some four years later, that company, Peruserver, has grown to a modest but dedicated staff of a half-dozen coders who share an office in the elite Miraflores section of Lima. Proudly, Ognio reflects on the developments. But he doesn't stress his own individual enterprise in such reflections or credit the developments to his own inner drive alone. As he explains: "That's something I personally owe in part to free software. . . . Being able to invent my own business, employ others, [and] get a bit of recognition among my colleagues . . . that has been an important goal achieved." For Ognio, playing with free software became a way of interrogating professional and personal goals and reflecting on the limits of education in programming under standard university training. But it also became a means to interrogate the future projected for him by family, academic advisers, and educational institutions alike. His frustration with university training and policies and his dean's warning that he not "dream" of becoming an advanced level programmer—one who might even join the development team of a large IT corporations such as Microsoft—were not experienced as recommendations to merely assent to. They become for him, rather, moments of apprehending a future that was prescribed but that he felt fundamentally disconnected from. Such moments of social misrecognition (Lacan 2002; Weber 1991) thus became opportunities to interrogate given futures. These processes of active interrogation around IT's given futures in Peru would indeed extend, too, to the countless hours of unpaid work he would later contribute as a prominent advocate for Peru's first FLOSS bill. Whether generating publicity around Proposition 1609 through the various interviews he provided national and international news outlets, organizing regional and global conferences on FLOSS in Latin America, or volunteering in less spotlighted roles as a flyer distributor or coordinator for Linux user group listservs, his activities spanned a range of commitments that pressed for broader social and institutional changes, and were rarely directly related to personal or professional gains alone.

Indeed such investments frequently drew him away from time he could otherwise have directly dedicated to his company alone. But he doesn't lament such investments of time, skill, or labor. By his account, FLOSS gave him what he called a "crystal ball" through which another future became visible for him. He thus narrated the time prior to his discovery of FLOSS as a period when he saw himself as "blind" to the other futures that were possible. Experimenting with free software became for him a mode of reworking limitations he came to see as being externally imposed. It was, in other words, not only the work of self-diagnosis that Ognio assumed within the space of play but also an exercise in social and institutional diagnosis and critique.

However much the process of information-engaged play yielded opportunities for an examination and reshaping of the self, though, it was an opportunity (or even demand) for an interrogation and reshaping of the external, social world that Ognio's account—like his investments in organizing for Peru's FLOSS bill—also stressed. His own case of personal transformation and achievement as a member of Peru's emergent information and "creative class," that is, was narrated not only to highlight his own sense of individual redemption and recovery after a process of pronounced self-exploration but to highlight the cultural logics that manifest within local systems of education, professional training institutions, and even private spheres that keep norms of educational and professional formation in place. Pressing for the interrogation of such norms, the playful information engagements of local FLOSS advocates similar to Ognio push forth the possibility of other futures than those established professional and educational practices reproduce. He would find, of course, that he was not alone.

Software Choice and Ethics of the Self

Fernando Gutierrez, a twenty-one-year-old student in systems engineering, played an indispensable role in the publicity efforts surrounding the

Peruvian free software legislation. He and several friends, all members of mostly the student-based Linux users' organization Pimiento, had built and maintained the web servers that globally distributed news and documentation around the Peruvian bill. They also volunteered their efforts to respond to the hundreds of free software supporters around the world who contacted the group in regards to the bill. And they later worked to coordinate the waves of supporters who in turn volunteered to translate the Spanish-language documentation for the Peruvian FLOSS bill into more than two dozen different languages.

Two years before these efforts began, however, Gutierrez was a second-year student in systems engineering who had just started experimenting with free software applications. He recalls in fact that it was another friend of his, Breno Colom, who eventually helped to found the Pimiento organization, who had first began toying with the Debian distribution of Linux, and who had then passed it on to his friends: "We became friends playing '"Dungeons and Dragons.' . . . His dad had an office near my house; we went there almost always to talk, to play with the computers. One day he told us that he had installed Linux on his computer; we tried it and he gave us the copies. We had always been interested in computers, of course, so it was like a challenge to us, to start with a new operating system. At first it was sort of a challenge to see who did this first, or who could make [this] work, or who configured what correctly. Little things, really more for fun."

Gutierrez recalls experimenting with several different versions of Linux in the next couple of months, trying Red Hat after Debian, then switching over to Slackware and finally settling on the Gentoo Linux distribution. His afternoons were often spent with his university friends tinkering with the various features of each distribution. Increasingly, however, his after-school experimentations with Linux and free software began feeding back onto reflections on the university's own technological capacities, affordances, and policies. As he recalled, "We began to see that at university, we didn't have much space for Linux, and all [their computers ran] Microsoft, so that using Linux for certain things became difficult . . . We saw it in only two or three courses . . . generally in the operating systems course, then around database server aspects, and web server ones. [But] we were very limited. Because out of the 30 or so computer labs, only two rooms had Linux or [an]other [non-Windows] operating system installed. The only active support the university offered around Linux was a bad mailing list."

Elaborating, Gutierrez described the technological affordances designed into a virtual classroom website his university launched. The site was intended to be a space for use by students and teachers, where assignments, materials, and discussions for classes could be digitally conducted and archived. As he explained, "We had to send homework assignments and that sort of thing through it, but it can only be accessed with [Microsoft's] Internet Explorer [IE] browser. We insisted the head of the project modify it so we could access it [using] Linux, or at least from [Linux's] Opera browser on Windows. But they [just told] us to use IE. They didn't give us any other option to choose from!" He recalls being one of five students who helped prepare a letter asking for the virtual class system to be redesigned to allow students who were using non-Windows operating systems and software to be able to have equal access and ease-of-use. They presented their letter and their case to the head of online services at the university but without successfully convincing him to alter the site. Reflecting on the response, Gutierrez emphasized the need for a larger internal, student-based mobilization to demand reform in university policy: "We would like to change things, but we . . . need more people in this movement in order to be taken seriously."

Such a realization motivated campus FLOSS users to begin organizing efforts to promote and educate the student body on free software's relevance to university students. The group began speaking to students in the campus halls and contributed articles to the university newspaper that argued for more free software support on campus. They also began organizing a larger conference around free software that would be targeted for university students. As Gutierrez recalled his impulse for the idea, "In that time, I had a friend who was the student representative for the faculty, so I asked her to talk with the dean about our project. I gave her all the documentation and asked her to discuss the use of the auditorium of the faculty for a conference [around free software]."

Although the permission was denied, Gutierrez and his fellow student activists remained undiscouraged by the administration's response, and the group started distributing copies of Linux CDs inside the university in the free software tradition of a "2 × 1" exchange. Under the model, a copy of a Linux CD could be exchanged without charge to a user for two blank CDs that could then be used to make new copies of Linux CDs and that would be redistributed to new students. Describing the effect the group had on the student body, he told me, "It's difficult but we have made progress. We might have been the only ones [who cared about free software] a year ago, but now we've gotten it so that there are more students that have at least tried Linux. And they started a pyramid in the way of distribution since they each had to teach two other students how to use Linux." When asked why he dedicated so much of his work and time to his advocacy for free software, Gutierrez emphasized a practical and political necessity to making technological choices visible to students. He insisted, "We learn more with

Linux, but we also have the *right* to use it. . . . It's firstly a matter of knowing the option of free software. . . . They are already using it in companies
where we are going to work in the future, so it's a thing we *must* learn. And
we are in university for that, to learn about the best options available. We
can't leave university and then, when our bosses tell us to maintain a Linux
server, tell them that we didn't learn how to do it."

He elaborated further on what he experienced as the difference between
"learning" on free software and closed, proprietary software systems such as
Microsoft's Windows, stressing again the centrality of technological choice
and the enhanced opportunities for technological learning with FLOSS.
Even when voluntarily coding with free software increased the difficulty of
class assignments, he recalled that such challenges seemed only to enhance
his ability to learn. He said, "The advantage of [free software] is learning
more than one way to do things. . . . Like in a theory of languages and com-
pilers assignment I had, I decided to develop it in Linux. It was ten times
more difficult than just using [Microsoft's] Visual Studio, but in the process
I learned ten times more things than the other students. My teacher didn't
see the effort, he just gave me a [standard grade] for my work, just like a lot
of other students. I spent much more time in it, but that didn't matter for
him. It seemed unfair but I [did] get a personal reward from it."

The experience of his university's lack of receptivity to an alternative technology, however, fostered Gutierrez's questioning of the administration's
authority over student education. He expressed such doubt, for instance,
when he told me that "generally, the majority of [the administration]
doesn't know a thing about Linux or free software, and they say that they
encourage the students to do research but the truth is that they don't. . . .
There are some teachers who congratulate their students for researching free
software on their own, of course. [But] I think that some teachers are afraid
of free software, just as people can be afraid of new technologies or new
ways of doing things when they are used to doing it only in one way." His
faith in the leadership of the university administration was only corroded
when, shortly after the introduction of the free software bill to the Peruvian Congress, it signed a service contact with Microsoft. By the terms of
the agreement, the university would integrate courses on programming for
Microsoft's Office Suite in exchange for discounted license fees. By Gutierrez's reading, the contact seemed designed to curtail free software education
in university classes: "The contract was not free for everyone to look at, but
. . . we think that they [used it to] try to stop the free software movement. . . .
It was like a cold shower for us. We were very happy about the coverage [the
free software bill was receiving], and really thought that we could win this
one, until the university's contract with Microsoft came through."

The actions of the university, however, only intensified Gutierrez's conviction that the free software–focused campaigns of student activists were necessary. Following the introduction of the FLOSS bill to the Peruvian Congress, he began postering campus walls with flyers designed by several of the Lima-based free software organizations, including PLUG, which voiced support of the bill. Other students, he recalled, volunteered to assist with the postering efforts after learning about the bill: "It was in the middle of the news climax when we realized that the people inside the university were talking about it, the poster efforts got their attention, it was the perfect time to begin more active work. . . . I don't agree with the politics of the university, signing exclusive contracts with Microsoft and closing the doors to Linux and free software in general. I think that free software option is part of the future of computers, and if we are going to work with computers all our lives we need to know about it. For me it's just incredible to imagine that a fourth-year student of systems engineering wouldn't know a thing about Linux."

Asked why he invested such intense efforts to promote the free software legislation within the university's campus, he replied, "The problem is that we wouldn't need a law if the people were conscious that they don't have to use proprietary software, that there is an option to that, but since they don't know about it they have to find solutions to their problems. So, in order to start using free software in the government and in the private companies we need professionals that can handle it. I regret to say that there is not one university in Peru that prepares students to use free software. So [I thought]: if teachers aren't teaching their students [this], maybe a student can start changing things himself. That was what made us decide to start the activism movement in favor of Linux inside the student body." Although playing with free software began as an exciting social activity shared between friends and as a personal challenge to master a new technical system for Gutierrez, his experiments came to be experienced as acts that dramatically enhanced his capacity to question and learn. Central to such an experience, he insisted, was the capacity to suddenly realize new technical choices, options, and possibilities when none had seemed present before.

Experiencing such freedom in his playful engagements with free software, however, revealed the presence of multiple barriers that continued to limit the possibilities of new modes and norms for information-based learning. If it was an unrestrained creativity that he was able to experience through his tinkering with free software, however, his university programming classes increasingly betrayed not merely a highly controlled and constrained approach to coding but also an intolerance to other modes of learning that limited the practical ability to learn new systems and that curtailed the possibility of even knowing distinct alternatives existed altogether.

Indeed, for Gutierrez, experimenting with free software became an experience that centered on more than exercising individual choice and that raised new questions of institutional ethics and policies. Once opening an opportunity for a new choice and alternative to be considered, however he'd grown in his conviction that it was precisely a choice that he and students like him could not afford *not* to make.[13] Convinced that one learned best when allowed the capacity to consider new alternatives and enact decisions around such alternatives, his university's lack of accommodation of technological options appeared to him to counteract the very mission and obligation to education. If prior to his experimentations with free software, then, the university's exclusive support of Windows and proprietary software systems could have been taken for granted as a decision intended to facilitate students' education, following his experience of tinkering with free software, such administrative policies became visible instead as ethically questionable ones that would inhibit the intellectual and profession development of students and prohibit precisely the possibility of their realizing their personal potentials.

To engage with free software play, then, allowed Gutierrez to explicitly confront the decision-making authority of the university administration and allowed him—within such an encounter—to interrogate the basis of its exercise of authority. Narrating his experience with free software was not merely a means for him to recount an experience of personal progress and intellectual development, but became a way of narrating his realization of the obligations of social institutions to fulfill their social duties. It was an experience, in other words, that would allow him to produce, with unwavering conviction, an insistence that students "learn more with Linux, [and] ... also have the *right* to use it." Personal transformation—in his own imagination, from an individual lacking technological and educational choice to one with fewer restrictions and from a person with limited capacity to learn to one with enhanced abilities—certainly figured into Gutierrez's experience of play with free software. But as much as his free software play compelled a process of identity transformation, so too did it compel something more. Exceeding the boundaries of self, the forces of play also triggered the will to reshape a prohibitive social world into one in which alternative modes of development and growth would be newly possible.

The Danger of Play

Victor Turner's exposition on states of liminality offered a crucial model for thinking through play as a transitional force. Examining rites of passage among the Ndembu tribe, Turner emphasized an intensity of energies that were enmeshed within the temporal and cultural space positioned between childhood and adulthood: "Liminality may perhaps be regarded as the Nay to all positive structural assertions, but in some sense the source of them all, and, more than that, as a realm of pure possibility whence novel configurations of ideas and relations may arise" (Turner 1967, 97). Saturated with new meaning and potential that exceeds the borders of either the identity that was once had or the identity in formation, the liminal actor would typically be regarded as a polluting entity (Douglas, 1966). By Turner's formulation, however, liminality exists as a uniquely permitted condition of disorder and denaturalization that would otherwise require stricter policing. Unless carefully controlled and vigilantly governed, the liminal being as a concentration of such disordering energy threatens to destabilize the realities that surround him or her.

Indeed, Turner makes explicit the profound disciplinary work that is required to manage liminal potentiality. He described the elaborate, theatrical rituals mounted to teach liminal actors to "see" appropriately when masked beings are summoned to provoke the liminal actor's contemplation of everyday relationships and features of his or her existing society usually taken for granted. Such masked actors mean to compel a new recognition of social forces as powers that generate and sustain individual subjects, at the same time acknowledging the deep ambivalence to a subject's actualized transformation. Although such rituals may be harnessed to affirm and strengthen socially prescribed identities, they also reveal the possibility of alternative responses. As Turner (1967) writes, "Liminality is the realm of primitive *hypothesis*, where there is a certain freedom to juggle with the factors of existence. . . . There is a promiscuous intermingling and juxtaposing of the categories of event, experience and knowledge, with a pedagogic intention. But this liberty has narrow limits. The neophytes return to a secular society with more alert faculties perhaps and enhanced knowledge of how things work, but they have to become once more subject to custom and law. . . . They are shown that ways of acting and thinking alternative to those laid down by the deities or ancestors are ultimately unworkable and may have disastrous consequences" (106). Turner draws attention thus to the simultaneous productivity and danger of play. Through it, an individual may emerge more prepared than ever to embrace a preassigned role. There remains, however, the potential that an individual—who, having newly encountered the realm of "hypothesis" and confronted alternative modes of being—may reject the prescribed future and choose otherwise.

It is such a creative, revelatory function of play that enables it to destabilize and upset established worlds. Made aware through their playful engagements with technology that alternative modes of ethical existence are indeed possible, Peruvian FLOSS users began to experience the projection of their intended futures by authorities as artificially restrictive, individually prohibitive, and even socially unethical. To explicitly adopt practices beyond the space of play, that aimed to reshape such restrictive social worlds, then, became a choice many were convinced they were obligated to make.

Understood as processes intertwined with the cultural generation of self and personal transformation, play becomes less about an undoing of labor or the culmination of the space of individual freedom and comes to be seen more as a zone of exploration where the self works to recognize and assess both existing and alternative worlds. Within play realms, actors may be given the allowance to interrogate the differences between internal and external worlds and desires, but pressure remains on actors to still come to accept such distinctions and also identify what transformations would be necessary to reconcile such differences. Within the realm of play, then, individuals are confronted with the possibility of self-adjustment to reconcile worlds but not without at once interrogating other potential paths as well.

Play, recognized as an intensely fraught space whose energies may be channeled for individual adaptation and the exploration of existing or other social worlds, then, pushes beyond the simple neoliberalist framings of play as labor's oppositional pair and digital labor as something other than the triumph of global "creative classes." That Peruvian FLOSS activists' initial engagements with free software as objects of playful, individual experimentation and interrogation could lead to collective forms of political action, then, demonstrates how they anchored their play activities with technology within frameworks of transition, transformation, and inquiry rather than within processes of diversion and the absence of labor. Indeed, embedded within FLOSS's play practices was not merely the capacity to recognize the real as a space that they should legitimately reshape their identities for but also a recognition of the real as a space itself in need of profound institutional and ethical reshaping. Key, too, however, is how their playful engagements with code were not understood as being exclusively or even primarily channeled into the enhancement of technology or the optimizing of production. More powerful, instead, was play's ability to expose the need for new, externally directed investments that aim toward a transformation of existing social norms, institutions, and policies in the name of making other worlds and information futures possible.

A work table shared by indigenous language and open technology activists at the software translation workshop held during the 2011 Sugarcamp Conference in Puno, Peru.

6 Digital Interrupt: Hacking Universalism at the Network's Edge

* in·ter·rupt

1. (v) to cause or make a break in the continuity or uniformity of (a course, process, condition).
2. (v) to cause a break or discontinuance; interfere with action or speech, especially by interjecting a remark.
3. (n) computer-related. A hardware signal that breaks the flow of program execution and transfers control to a predetermined storage location so that another procedure can be followed or a new operation carried out.

The last time the better part of the team of a dozen or so young, transnational techno-activists gathered together for a project collaboration they had worked from the wired office space of Escuelab in downtown Lima. Founded in 2008 as an alternative learning and research center bridging "technology, art, and society," Escuelab[1] in its short existence had managed to build a network of self-identified media activists and free software–centered technology hackers—from across Latin America, Europe, and the United States—who converged, virtually and otherwise, for various project collaborations. On this particular afternoon in April 2011, however, they reunited on the top floor of the Municipality of Puno—the capital of Peru's southernmost province known more as a center of indigenous culture and the nation's "folklore" capital than for its role as a site of high-tech advocacy. Having traveled from points across the world including Barcelona, La Paz, Lima, and the local towns surround Puno, the participants came prepared with their own equipment so that the municipality's offline meeting room could be converted into a digital workshop of sorts.

Each carried a laptop—some came with two—many displaying cases covered with stickers printed with the bold logos of free software applications and slogans favored among global information activists. Several models of green and white XO laptops—the MIT-derived, child-centered computers that are the center of the global OLPC program that Peru has famously become the largest partner nation with, were also scattered around the table. A few wireless USB routers, a portable projector, chargers, and digital cameras rounded out the collective stock of gadgets. Such items were splayed out on the large conference table as the temporary workspace began to be assembled and as participants settled in with a half-dozen other volunteers to begin a two-day long project to translate software for the XO laptops into the indigenous languages of Quechua and Aymara. Although the project was still a long way from completion, the hope was that the results could be installed onto the nearly half-million XOs (a number that was expected to increase to 850,000 by July 2011) the state had distributed to public schools across the country.

Laptops screens were just being opened when Francisco Ancco Rodriguez—a retired public school teacher from Acora, native Aymara speaker, and now volunteer translator with the session—called for the attention of the party. Drawing out a round bundle of brightly colored woven cloth, he unfolded it onto the table's surface and spread the cluster of dried coca leaves that lay at its center evenly over the fabric bed. Sliding his hands over the thin slivers of green, he selected three and arranged them into a small fan formation. As participants then began to bow their heads one by one in slow succession, he began leading a short prayer in Aymara, holding the fan of coca leaves firmly in front of him all the while. Then, before taking the portion of the medicinal plant to his lips, he invited the group to draw their own shares from the common collection.

The ritual, a ceremonial offering known in Andean communities as a *despacho,* is one that the local participants from Puno had seen before— although not one that most of the international technology activists were accustomed to opening their own events with. Performed within Andean communities prior to many auspicious occasions, it is still understood today as a means of communicating with and honoring the spirits of the natural world, representing a triangulated union between the worlds of man (Kay Pacha in Quechua or Aka Pacha in Aymara), the world above (Hanan Pacha or Alaj Pacha), and that below (Uk'u Pacha or Manqha Pacha). Most commonly, it is framed by practitioners in a simplified version as an offering to the Pachamama—the expression of female cosmic energy typically translated as Mother Nature but in the Andean world associated as much with death and blood as with birth and life or to the Apu mountain spirits. In the realm of anthropological studies, however, the ritual has been most attended to in its offerings to the devil, el Tio, in ceremonies miners enact before descending into the depth of mine caverns (Harris 1995; Nash 1993; Taussig 2010).

In whatever instantiation, however, it is practiced as a means of entering into a careful and cautionary dialogue with ambivalent forces that are understood as not fully controlled by man, forces that might take or destroy life and its productive energies as much as it might give or feed such energies. Such powerful and potentialized "devil's energy," as Andean anthropologist Olivia Harris (1995) explains it, is one that is still understood to run through varied objects in the Andean world vision, encompassing "a whole spectrum of sacred beings: the mountains, the dead; powerful, untamed places such as gullies and waterfalls . . . as well as the Tio and Pachamama. The defining character of these devils is not so much evil or malice as abundance, chaos, and hunger. . . . [They] are the source of both fertility and wealth, and of sickness, misfortune, and death. They are unpredictable and very powerful" (312).

Such cautionary tones that underscore the ambivalent forces surrounding technological development are indeed unusual ones to be invoked within the global circuits of digital production. Better marked by insistently hopeful or emphatically celebratory tones that assert a certainty of innovation potential and assurance around the productivity of the collaborative partnerships that network varied research- and commercially oriented spaces together, high-tech's grammar is one that has typically left little space for cautionary (let alone critical) reflection. And perhaps nowhere has the ability of such a language to effectively evangelize the unquestionably empowering impacts of technological distribution—for users and producers alike—been stronger than around cases of emerging ICTs, including the OLPC.

Indeed, given Puno's own unique activities around ICT-based developments—one that could unexpectedly draw in a surprising range of participants from across an array of global sites to the southern peripheries of Peru—it seemed a likely site from which high-tech's future-enthused fervor might newly echo. Since local deployments around the OLPC project began in the province more than four years ago, Puno had steadily begun to earn unexpected acclaim within certain global technology circles for its experiments with the XO laptop. This, despite the fact that most of its activities lacked official endorsement from either the Peruvian Ministry of Education or the OLPC Foundation in Cambridge, Massachusetts. Nonetheless, it was from Puno's highland stretches that the first XO-based user manual—a one-hundred-page, teacher-centered text later distributed online and translated from Spanish into English, French, and Arabic (Salas 2009)—was written and published. It was efforts from Puno, too, that organized some of the largest rurally centered conferences for XO users—the latest of which drew hundreds of participants from across the province and from across the larger, global XO developers' community. And it was there that translations of the XO interface into Quechua and Aymara aimed to be among the first in indigenous languages.

There is, however, little sense of any pitched elation or exhilaration for having gained such global distinction expressed by the participants of the translation workshop. Neither do they emit a collective confidence around what future promise technological change could bring to a "peripheral" site such as Puno or what its inclusion into global networks of high-tech collectives and with "distant" centers of innovation might deliver. Standing as they are around the municipal conference table with heads quietly bowed and attention turned to themes of technological ambivalence, there seemed to be instead a consciously deliberative reflection effected—one that broke with the enthused tones summoned by high-tech's global development networks and that purposefully called forth a different sort of presence from participants instead. Theirs was a moment enacted as a form of digital interrupt, one that, similar to that in speech, caused a discontinuity in the ongoing action and interference in the usual activity of the larger network. Or, as in computing culture, one that provided a signal whose emission could break the flow of a program's execution, enabling a transfer of control so that a new operation might be carried out instead. In both cases, it is a simple event of no more than a modestly posed interjection that nonetheless creates an opportunity to catalyze a larger rupture, unsettling what had been allowed to operate before—without justification or explanation—as expected procedure.

Such an act of interrupt calls forth, too, something akin to the "epistemology of the sacred" Isabelle Stengers (2007) has invoked as key to what she frames as a cosmopolitical critique. Seeking to unsettle the "unifying referents" and shared givens that have long fastened modernity's dominant frameworks, the participants' interrupt intervenes around the practices of the larger innovation network undertaken in the name of collaboration to question what provisions underlie their assemblages. Underscored, that is, is a tension that for however impressively varied the institutions and actors involved in high-tech collaborations may be, and for however expansive the shared intellectual, economic, and social resources channeled, uncertainty and instability can still persist. And with whatever convincing strength they could all appear to hold together to affect a sense of "pacific collaboration," concerns around how such a veneer of consensus could emerge still abound. Around the translators' table of mixed tools and resources, then, the varied "unknowns" that surround hi-tech's productive collaborations could not help but "resonate" (Stengers 2007, 76).

And perhaps with good reason. While Puno's recent OLPC-based developments saw growing interests drawn together—with national government ministries, US universities, transnational research and innovation centers, and individual Western-based hackers building new investments into the project and offering to channel resources into rural zones for it—on Puno's streets, local unrest was brewing. Just days after the translation workshop's came to a close, the area would erupt with regionwide protests involving thousands of indigenous community members from across the province (BBC 2011; Cardenas 2011; Guerrero 2011). Foremost among the reclamations was the national government's growing land concessions to multinational mining companies and its persistent disconnection from the interests and concerns of rural communities and local governments alike. Only months earlier, Puno had been a key site of nationwide protests in which sixty thousand public school teachers demonstrated against a new law aiming to nationalize teaching standards in public schools—and which many read as evidence of Lima's disinterest in the daily realities faced by provincially based citizens and workers. During protests, Puno's airport was set afire, roads were blocked, and cars overturned (Carroll 2007; Living in Peru 2007; Salazar 2007; Uco 2007). Although rarely mentioned or even acknowledged by OLPC advocates abroad, incidents of pitched political conflict—involving educators and community leaders—were erupting in local XO deployment zones in Peru. And even as such local realities remained consistently neglected abroad, local actors working around OLPC's project continued to pose the question of technology's role in such dramas—asking how the promise of networked partnerships and the security of peaceful collaboration came to be such convincing "givens" to begin with.

Peace, Love, and Technological Universalism?

In the vocabulary of contemporary innovation-focused cultures, caution has not often been highlighted for playing a central role. Innovation language has been better known for its insistently hopeful, forward-looking, and emphatically celebratory tones, shared and echoed by a networks of actors (Callon, 1986; Latour and Callon 1981; Latour 1987, 1990, 2004a) distributed across complexes of research- and commercially oriented spaces. Daring to strive—collectively and faithfully—toward realizing versions of technology that press the limits of the novel and risk-defying, innovation's network of actors unite under their shared commitment to the idea that their leaps of faith forward can bring forth progress-oriented change. It is, as Kaushik Sunder Rajan (2007) stresses in his work on biotechnology's global networks of innovation, a mobilizing language fundamentally animated around speculative visions and promissory hype: "a game constantly played in the future to generate the present that enables the future" (34). Relentlessly assured of the possibilities of the improved future and its continual realization through the pursuit of the next "new, new thing" (Lewis 2001), innovation talk minimizes potential error and the possibility of itself as anything but predictive—promising instead a salvation from a present continually redefined as one of constraint and limited possibility. Particularly in the developing world, such framings of the constrained present—and the need to catch up with a "present" others are already leaving behind—can prove especially resonant (Escobar 1994; Ferguson 1994). Unanchored from any particular territory, then, innovation talk emerges as a profoundly affect-laden practice whose deeply constitutive power is captured in its ability to generate and mobilize varied forms of work across expanses of lab, market, and consumer spaces—including those, indeed, well beyond the boundaries of the West.

It is a language evident, too, in the institutional efforts that surround global deployments of the OLPC project that have promoted—as one ad for the XO in Peru simply and plainly states—the message of "technology as the key to the future." In OLPC ads sponsored by Peru's Ministry of Education for its XO deployments, in which such slogans have appeared, nothing more than a single XO laptop is featured against a stark, green background with the simple, seven-word declaration printed beneath—calling attention more to what is not said rather than what is. Notable, after all, is the absence of specification of the local contexts of the XOs' deployment or the particular users whose futures will be affected by them. More than exhibiting an underlying logic of technological determinism and an assertion of technologies, generally, as drivers of history (Smith and Marx 1994; Winner 1989), such unconstrained assertions around a single technology betray a sense of technological universalism—one in which the same technology (whether it be the Internet, a mobile phone, or, as in this case, a globally purposed laptop) can serve as a multifunctional, universally common solution for all users. Or as the Peruvian education scholar Eduardo Villanueva describes OLPC's globally ubiquitous approach to technology: "OLPC still believes in the power of one technological solution transforming realities as varied as Afghanistan or Uruguay" (Villanueva 2011).

Such a framing of technology echoes much of what American studies scholars Siva Vaidhyanathan (2006) and Carolyn de la Pena (2006) have characterized as a "techno-fundamentalism" held among contemporary designers of new technologies, who insist that any human problem—even those caused by technologies themselves—are best solved by newer ones. But while the former discourse operates by instilling a faith in varied technologies to present solutions in given social or cultural contexts, in the latter, it is a single technology that is framed as a flexibly variant, global solution for a broad range of existing problems. Or as Negroponte explains the unbounded promise of the XO to global technology developers and policy makers: "Access to a connected laptop or tablet is the fastest way to enable universal learning" (2010b).

It is, of course, poverty and underdevelopment as persisting global "problems" that technology is summoned to provide the key for in ICT for development (ICT4D) projects such as the OLPC. The rapid international spread of various ICT4D projects—OLPC most prominently among them—as well as the complex global coordinations and commitments they have orchestrated between broad populations of technology designers and deployers evidence the eagerness that surrounds the notion of technological universalism today. Lars Bo Andersen points to the diverse range of entities whose activities are orchestrated for the OLPC project alone: these include representatives from the MIT Media Lab, including Negroponte himself; actors from US and Taiwanese corporations such as AMD, Brightstar, Google, Red Hat, and Quanta Computer; strategic partners such as the United Nations Development Programme (UNDP); and ministries of education from partner nations, transnational and locally sited NGOs, and local schools themselves (Andersen 2011). Drawing together such varied, multisectoral interests—which readily cross scales of the international, national, and regional—ICT4D projects organize via globally distributed network of actors. Their coordination presents technological universalism less as argument or belief and more as its own self-evident fact (Latour 2010; Stengers 2007, 2010) One, moreover, that, in a post-Internet age that touts the coming of the seamless network and ubiquitous computing, obligates "conversion" (Stengers 2007) and acceptance from parties in order for the "future's key" and the ever-growing promise of global connection to be accessed. In the meantime, nonbelievers or skeptics can be framed as the other to progress advocates—willfully underinformed, backward, or simply disconnected (Latour 2010).

The "obligation to be converted" by the universalizing promise of modern reforms, however, is one of the cosmopolitical orientations Isabelle Stengers (2007) asks us to reconsider. Or rather, it is the objectives of "peace" and "peaceful consensus" as allegedly known ends for which "unifying referents cause conversion" (80) that she asks us to return to. Under such modernist modes of intervention and reform—from Western science to modern political and liberal economic norms—she notes that answers can too frequently present themselves as universalizable, even before a question can be posed. To interrupt the "answer" technological expansion naturalizes, however, is to expose how tenuously its "peace" and consensus lay just beneath the surface. It is to highlight, for instance, the vast differentials of power and resources of numerous kinds—social, economic, cultural, and material—that remain between the diverse actors whose efforts must be networked to sustain ICT projects and whose commitments must be secured in order for the self-evident answer of "technological extension" to emerge as it does today. It is to bring forth, then, the possibility of a vast expanse of other questions rarely asked when the technological answer is asserted as an "obvious" consensus: what happens in the process of fabricating such stabilizations? Why would such conditions be required for "peace" to exist? What should one make of the unrest implicated by the "conversion" of actors—or what's lost, forgotten, and undone in the meanwhile? And why should there be such high stakes to relinquish the unknown at all?

As it turns out, the periphery becomes an ideal place to pause and undertake such inquiries.

Collaborative Technologies and National Technologies

A visit to the Peruvian Ministry of Education's (MINEDU) website[2] for its national OLPC initiative makes plain, for even a first-time visitor, its vision of technological universalism. After entering the site via a page that features a single profile of a young student with her gaze fixed into the screen of her XO laptop, visitors can click onto a "testimonials" link that leads to another page compiling the accounts of several other newly minted XO users. The page features a selection of nine photos of rural students with quotes—labeled as *testimonials,* although none is attributed to any named speaker—alongside each. Beside one photo of two girls seated on a grassy slope with an XO atop one of their laps and a stray goat grazing in the background appears the quotation, "Today we dream of a future, tomorrow, we'll achieve it." Beside another, where a student sits in a classroom typing into her XO, appears another quote, "Today I discover, tomorrow I'll innovate." And beside the last image in the column of "testimonials," a photo of a group of seven smiling XO owners with the green and white laptops raised over their heads, appears the quote: "Today we have our OLPC, tomorrow we'll be prepared for the future." The page of anonymous quotations supplies no further information on where the photos were taken or when quotations were recorded—a curiosity, given the program's heavy financial investments. According to reports, some $82 million USD ($80 million on hardware and a notably minimal $2 million on teacher training) was spent to purchase the hundreds of thousands of XO laptops for elementary school students across the nation (Talbot 2008).

The message of fostering changes that would allow rural students to access a "future" they otherwise wouldn't have is one the ministry has championed since it first began deployments of the OLPC program in 2007. Back then, the project had as a small pilot program in the mountain town of Arahuay, a rural village of some 740 residents four hours outside of Lima, where most families farm—avocados and cherimoya among other agricultural goods—for their livelihoods. Although any number of pilot sites might have been chosen, the ministry emphasized its interest in prioritizing the nation's most disadvantaged students to ensure that they would be "equally" granted the same "quality" educational opportunities other students could count on. As explained on its website, "The OLPC Program responds to the demand for educational quality and equity through the integration of information and communication technologies (ICTs) in educational processes . . . especially in areas with the highest poverty, high illiteracy, social exclusion, population dispersal and low rates of concentration of student population, to contribute to educational equity in rural areas. The program seeks to improve the quality of education, for which it is necessary to modernize and strengthen the role of teachers."

Oscar Becerra, the Ministry of Education chief educational technology officer and the head of its OLPC deployment, spoke to a reporter with *MIT Technology Review* in mid-2008 in the months before the pilot program was expected to expand exponentially to one of national-scale deployment. Although no official studies had been completed to demonstrate the result and effects of the pilot program, and although many early press accounts on the project stressed that it was still unclear what benefits the program had achieved or what it was expected to result in (*Economist* 2008; Paul 2008, 2009; Vota 2009), a rapid expansion was still underway. By the end of the following year, some four hundred thousand new XOs were to be distributed to nine thousand rural schools across the country. And Becerra, who had earned a doctorate in psychology and bachelor's degree in physics, stressed that the project expansion would allow rural children the freedom to not just imagine their own "future"—but to choose one for themselves: "Our hope for [that student] is that he will have hope. We are giving them the chance to look for a different future. . . . These children who didn't have any expectation about life, other than to become farmers, now can think about being engineers, designing computers, being teachers—as any other child should, worldwide" (Talbot 2008). Becerra, who had worked for twenty-two years at IBM, serving as their Latin America education segment manager before joining the ministry, was well practiced at connecting the local, particular ambitions of children of rural Peruvian communities to the universalizable ambition "any other child" might have. And he bridged the shared problems confronted between local and global scales in speaking to foreign journalists again when he described a "global crisis in education" as catalyzing demands for projects such as OLPC's worldwide (Bakk-Hansen 2008).

Such global linkages notwithstanding, however, there's no doubt that problems particular to the nation were also motivators for Peru's considerable investments into the OLPC program, and its commitment to deploying more XOs than any other partner nation (despite, again, the lack of data around its use in either large or small-scale deployments). Ministry of Education officials readily cited Peru's second-to-last ranking in primary school–level math and science education—130th place out of 131 countries—in a survey completed by the World Economic Forum in 2007 when explaining the necessity to seek out rapid and immediate education-based solutions like the XO. The early years of OLPC's pilot program also coincided with the ministry's heavy promotion of a highly controversial law, the "Teacher Career Law" (*Ley de Carrera Pública Magisterial*, one of the much-lauded educational achievements of President Alan Garcia's outgoing administration) that would base teacher promotions and retentions on national exam scores. And although state officials were initially careful to avoid explicitly pointing the finger at the 320,000 public school teachers for the nation's low educational ranking or for producing the "crisis" that required investments in programs such as OLPC—some 60,000 of them were already taking to the streets in protest against the law that halted schools nationwide for weeks in 2007. And ministry officials became considerably less discreet following the law's passage (*Andean Currents* 2007; Peru Support Group 2007).

Becerra in recent interviews pointedly referenced the shared objectives between Peru's OLPC project and the Teacher Career Law. In an OLPC-sponsored supplement published in the national newspaper *El Comercio* in January 2011, he stressed the stakes involved for Peru's global standing and underscored the notion that there was no time to waste: "The low quality that public education has [in Peru is known]. . . . The fundamental problem is that an education system can't improve if the teachers don't. The leading countries in education have teachers that perform best on international exams. This is what Peru will improve via its new Teacher Career Law. But this process will take 10 years. And in the meantime, what will we do with the children" (*El Comercio* 2011)?

Such framings of Peru's 350,000 public school teachers as the primary obstacle to educational reform and the very global futures of its children circulated prior to the XO's emergence onto the national education scene. The emphasis on the responsibility of individual teachers to continually improve their skills and the heightened public scrutiny teachers became subject to because of it were outcomes those working in rural provinces had become sensitive to by the time thousands of XOs began to be filtered into rural zones. And they were among the critiques the national teachers' union, SUTEP, charged the state with when the month's worth of protests began to be organized across the country against the national Teachers Career Law in 2007.

As the Ministry of Education went forth distributing hundreds of thousands of XOs across the country, however, public school teachers began to actively inquire whether it was the ministry's own preparedness around technology that now invited evaluation. Although laptops were arriving into the provinces and there were assurances training seminars run by the ministry would be organized soon after, the XOs came with few other materials, instructions, or guides as to how they should be integrated into rural classrooms or local curricula. Wondering how national and local governments planned to redefine their roles and presence across rural provinces—and particularly in rural communities—to prepare teachers, students, and families in integrating such new digital tools into community life, local teachers began to organize events to address curricular planning, content, and teacher training related to the XO. Public school teachers from Puno extended the initial invitation to the Escuelab technologists and media artists to aid in their planning of three technology-centered workshops—each organized independently of the National Ministry of Education—in 2009, 2010, and 2011. The 2011 workshop drew some four hundred attendees for a weeklong event in which presentations were given by teachers, technologists, and Escuelab participants—from Peru, Bolivia, Argentina, Ecuador, Spain, Denmark, and the United States—on topics including indigenous film in education, community-based web streaming, and open source software for the XO.[3]

Among the speakers from Puno for the event was Eleazar Pacho, a twenty-seven-year-old principal and teacher in a small public school in Lacachi, a village to the south of Puno, whose forty-two students were among the region's early recipients of XO computers in 2009. Pacho and his three fellow teachers at Lacachi traveled nearly four hours daily on local roads to arrive at and return from their village school from Puno. Accessible primarily though a handful of local combi buses that passed through dirt roads close by, Lacachi nonetheless earned a reputation for being one of the more successful regional deployments of the XO, where teachers and students maintained routine use of the laptops in class long after many regional schools had found the lack of institutional support unsustainable. It was for that reason that it was selected as the school that would partner with Escuelab's media artists for a project that was conducted simultaneously with the workshop activities in Puno, built around digital media and indigenous youth expression. And the audience, made up largely of fellow teachers from schools across the province, with local government officials and foreign IT company representatives scattered throughout, was keen on hearing his story.

Speaking to the auditorium, Pacho recalled the surprising arrival of the region's share of laptops and how they had appeared without prior warning, discussion, or debate. As he described it, "When they arrived, there was no other option [other than to accept them]. . . . Even though when they gave you the computer, it was really another duty on top of all the [routine] functions that teachers have, and we had never been trained to teach with such tools before." He stressed the curious certainty with which the XOs were deployed—despite the lack of preparation, training, or prior discussion and debate around the new tools—and asked his audience to consider with him whether such certainty was warranted: "I don't want to say it was bad idea. . . . We were told that the XO was a solution for education. [But] I think they should have reframed it, because there have been so many new interventions from the government in education, and until now, it hasn't gotten better."

He recounted, nonetheless, the vivid excitement he and his fellow teachers from Lacachi shared as they received the news they would each get their own XO and that the Ministry of Education would host a weeklong training session for it: "I remember the day that they were going to give us the XOs, which was a Monday just before our training on them would begin. And the three other teachers [from Lacachi] and I were very excited. We didn't have our own laptops back then, in 2009. [And] we were so excited that the first thing we [wanted to do with them was to] open them up and study them, so we'd be prepared for the upcoming trainings." Pacho recalled how he immediately began searching for XO user manuals online that he could share with his colleagues and he remembered with pride how he and his colleagues started experimenting with the XO's chat function before the training workshop began: "We started to test it out between the three of us [almost immediately] . . . [so that] when we got to the classes [the Ministry of Education had arranged], we could already start chatting between ourselves. I remember [attending the training, and] when lunchtime arrived, one colleague told me so via chat that we should leave class now to meet for lunch. And I responded, 'already?' So we could already share this type of communication at the beginning [of the Ministry's training effort]."

He stressed in his presentation the various programs and uses for the XOs in elementary school teaching that were already being explored at Lacachi. And he wondered aloud why, given the state's expansive investments in the program, more resources to train teachers, develop pedagogical tools, and coordinate the extension and distribution of such tools hadn't also been reserved. "With all respect to the organizers [of the OLPC deployments], we haven't had sufficient training around the XO. [When the laptops arrived] in 2009, we had a weeklong training session that was very productive, but since then, at the institutional level and at the generalized level in Puno, there hasn't been another." He added that many teachers were resourcefully innovating new applications, uses, and materials for the XO in their classrooms but that such commendable efforts seemed to have been almost entirely overlooked. And he wondered, too, why more wasn't being done to coordinate the work of such teachers—as if end goals had been met in simply distributing the laptops: "I know that there are many teachers that manage these XOs very well, people with very successful experiences using them. I think these people should be found and brought together—not just to tell them, 'get together so you can finish this job for me,' like one more duty—but rather with some kind of incentive for them—more training, or a certificate, or a plan, perhaps . . . so that they can form a core nucleus and help others in other contexts."

Despite Pacho's critiques of the incompleteness in deployment approaches, Lacachi's work around the XO to date still evidences multiple unique accounts of what its working case of rural deployment in Peru has produced. Surrounded by the green highland slopes of Puno's altiplano, with the closest vehicle-passable road still a half-kilometer and river's cross by foot away, Lacachi's school typically hadn't received many outside visitors prior to the XOs' arrival. Still, it was easy to see that its forty-two students had spent considerable time using their XOs since. On a visit to the school in April 2011, students could be seen walking to class, each toting their own laptop and some playing mp3 files from them as they traveled. Laptop hard drives were filled with older digital photos and student-produced digital archives taken for earlier class assignments, and students' parents dropped by classrooms to see how the new technologies were being used. Notably, however, Pacho used little of his stage time during the Sugarcamp Conference to stress such achievements. His presentation instead emphasized the growing body of still-unanswered questions that that the state's distribution of XOs and general framing of the OLPC project generated—and the means by which ordinary conditions faced by teachers, students, and local communities still appeared to be broadly underconsidered, however much the ministry's deployments stressed the use of new technologies as universalizable solutions to education's "global" problem.

His careful framings and gentle apologies to his audience and to ministry officials for "perhaps" offending them, in particular, betray an awareness that what he shares interrupts the dominant message the state would wish to be echoed by local partner schools. If he was concerned with who his queries might put off, what interests in his audience he might unsettle or mark himself as alien to, or how his presentation may disturb the image of consensus sought by the state, he doesn't say so, and it doesn't stop him. When he takes a seat after his presentation, he does so beside me. And he leans over, only to tell me, "That was my first talk in front of a large audience."

"Do Nothing" or "Partner": Preaching Global Conversion

From its beginning, OLPC was, of course, not a modestly promoted project. Nicholas Negroponte, the chairman of the OLPC Foundation, who had already earned his fame as the founder of MIT's Media Lab and initial investor in *Wired* magazine when he launched it, made his global vision and pointedly missionary zeal for the project readily known. (Fouche 2012) At the XO's debut at the World Information Society Summit in Tunis in 2005, he boldly predicted that some 100 to 150 million XOs would be distributed across the world by 2008. Endorsing such expansive ambitions, Kofi Annan hailed the machine as "inspiring" and "an expression of global solidarity" (Twist 2005)—praise that made instant international headlines after Annan demonstrated the XO's capacity to be powered by a wind-up crank before a rapt audience of international policy makers.

Such unapologetically immodest claims, fed by Negroponte's characteristically unabashed techno-utopianism, invited considerable scrutiny and (arguably much deserved) criticism when the XO failed to achieve its own elevated expectations. At the end of 2010, only two million units had been distributed (Negroponte 2010b). Although they were implemented across forty different nations, only two countries—Peru and Uruguay—had committed to more than 250,000 units, the minimum initially required by the foundation (Andersen 2011; Warschauer and Ames 2010). And only Peru—who intended to have more than eight hundred thousand units in circulation by 2011—came close to the one million units the foundation originally demanded as the minimum commitment when the project launched in 2005.

Explanations for the project's sobering performance began to percolate up (Dendorfer 2010; Paul 2010; Santiago, Severin, Cristia, Ibarrarán, Thompson, and Cueto 2010; Toyama 2010, 2011; Villanueva 2011; Warschauer and Ames 2010). Considerable attention was given of the failure to meet the much-touted $100 USD benchmark, the unreliability and poor design of some hardware features (including the mouse, key pad, and hand crank), and the US-based OLPC Foundation's inability to work earnestly with national government partners. Academic studies cited Negroponte's unflinching "techno-determinism" (Toyama 2010, 2011; Warschauer and Ames 2010) as resulting in a refusal to learn from lessons local implementations offered. A long-time adherent to Seymour Papert's constructionist model of learning, which viewed children as innately able to teach themselves, Negroponte insisted that computers could allow all children to self-educate without being hampered by schools or teachers and dismissed the need to invest time or resources into the formal monitoring and evaluation of OLPC deployments.

Speaking at a September 2009 forum sponsored by the Interamerican Development Bank, Negroponte told an audience of government planners and policy developers, for instance, "I'd like you to imagine that I told you 'I have a technology that is going to change the quality of life.' And then I tell you, 'Really the right thing to do is to set up a pilot project to test my technology. And [that] once the pilot has been running for some [time], to go and measure very carefully the benefits.' . . . This all is very reasonable until I tell you the technology is electricity, and you say 'Wait, you don't have to do that.' But you don't have to do that with laptops and learning either. The fact that somebody in the room would say the impact is unclear is to me amazing—unbelievably amazing. . . . There's only one question on the table and that's how to afford it? . . . There is no other question" (Negroponte 2009; Warschauer and Ames 2010). Negroponte too remained unwavering a year later when he reacted to criticism published in the *Boston Review* around ICT4D projects including OLPC. Sounding impatient at the continued demands for more comprehensive data to support his claims, he wrote, "In OLPC's view, children are not just objects of teaching, but agents of change. Many of our kids teach their parents how to read and write. I have no better story to tell. . . . OLPC triggers communitywide capacity building. Laptops arrive, and generators-for-hire appear, or suddenly, as in Rwanda, the school is electrified. In Peru and Paraguay, local, independent software developers and repair shops start popping up. Laptops get children, their families, their teachers, and governments thinking on new trajectories—every day" (Negroponte 2010a).

And, indeed, in recent months, the missionary resolve—and public receptivity—surrounding OLPC has only seemed to intensify, the emerging critique notwithstanding. Speaking at the 2011 TEDx Conference in Rio de Janeiro to an auditorium full of some eight hundred people and with another seven thousand watching live online, Rodrigo Arboleda, the Colombian-born president of the OLPC Association (and former classmate of Negroponte at MIT's School of Architecture), underscored the urgency for addressing what he called a global "epidemic" of educational underpreparedness (Klein 2011). As black-and-white images of ailing patients in hospital beds flashed on the towering screen behind him, he lauded OLPC as a "vaccination against ignorance," citing the former-president of Uruguay Tabare Varquez's original reference to the XO that framed the nationwide deployment there. Then praising the audacity of the Uruguayan program to distribute five hundred thousand XO laptops, one to every elementary school student in the nation, Arboleda (2011) issued a global "challenge to governments around the world to put laptops in the hands of every child around the world," declaring, "the resources are there. What is needed is the political will to do it."

That OLPC's leaders maintain an unshaken conviction around the project's extension, despite its own slower-than-projected growth, warranted consideration. Although the number of government partnerships it has secured to date is a mere fraction of its original goals, the number of nongovernmental partnerships it has fostered in the six years since its original launch, along with the growth of the volunteer-based software and hardware developers' communities around the XO, have been considerably more impressive. OLPC communities focused on education deployments have now expanded across six continents in eleven different languages, with some thirty-five collaborations established with educational institutions and universities in the United States—many of which stewarded modest international deployments of XO-based projects under volunteer-centered OLPC-corps programs. (These include a University of Pennsylvania–launched project in Cameroon, a UC Berkeley project in Uganda, and a Harvard and MIT joint project in Namibia.) Advocates have also been behind the launch of sixteen nation-specific websites (under names such as OLPC Brazil and OLPC France) to organize country-specific efforts (representing roughly 50 percent of the total thirty-four nations in which XOs are distributed), eight distinct pedagogically focused "learning community" websites, and four independent OLPC-centered news websites.

The volunteer-based communities for software and hardware development, likewise, have seen impressive proliferation. Overlapping significantly with established communities that grew around various FLOSS educational applications installed on all XOs, more than 110 distinct coders have contributed to the XO's "Sugar" operating system and hundreds more have helped to begin translations of its interface into over 120 different languages. Such enthused dedication was visible during the recent EduJam Conference[4]—which focused on Sugar software development and drew more than eighty volunteers from over a dozen different nations across the Americas, Europe, Asia, and Africa—for the event in Uruguay's capital of Montevideo in May 2011. Although few public school teachers were present and none gave talks, other "celebrity" developers in the XO community—including MIT Media Lab director Walter Bender and OLPC's lead software engineer Chris Ball—were featured as keynote speakers during the three-day-long event. Such abilities to organize efforts and extend international collaborative partnerships are not only a central part of what's enabled OLPC to achieve the global visibility and momentum it has—but are, according to OLPC representatives themselves, key to the project's productivity and its ability to treat an "epidemic of ignorance." As Arboleda (2011) said to the audience he drew of nearly eight thousand in Rio[5]: "We can do nothing, or we can create public sector, private sector and NGO partnerships—and we can distribute [XOs] in massive quantities."

The persistently persuasive work of such framings of the OLPC and the urgency of its technologically salvational mission, however, is one its critics underscore (Morozov 2010a, 2010b; Toyama 2010, 2011). Given its broad ability to recruit new project participants—and the growing investments of labor and resources that are drawn together across diverse global sites to treat an ignorance "epidemic"—the "technological utopianism" of the OLPC solution, as Evgeny Morozov (2010b) writes, seems almost certainly "here to stay." Indeed, its framing of the field of possible action as either to "do nothing" or to "create partnerships" is an appeal issued through OLPC's network that local actors and project participants are aware they've been subject to.

Sdenka Salas, a public school teacher from Puno and one of the key organizers of the teacher-centered workshops for the XO in 2010 and 2011, recalls the certainty with which OLPC's reforms were predicted to bring to local schools across the province some years ago. Salas herself had earned some degree of renown in the past several years in certain international educational programming communities, largely through her own volunteer efforts around the XO. Her work to build localized technology projects in Puno, however, had actually begun before the province first received some eight thousand XOs in 2010. Several years earlier, she helped to form a core team of local teachers, student FLOSS activists, and public sector workers, who volunteered to help build free software–based computer centers in local schools (three of the largest public schools by the end of their efforts). With the arrival of the XOs, they began to center their work on localizing OLPC to the cultural context of the region, organizing youth in the area to assist in Quechua and Aymara translations of the XO software and interface, and organizing the first rurally based conference for volunteers, which drew a global community of hackers and technologists, as well as several hundred public school teachers from across the region, to Puno in 2010.

At the request of the local ministry of education, Salas took on an unpaid project to compose a user's manual for the XO, targeted for local teachers. The resulting one-hundred-page book, compactly printed with the image of a young boy in an Andean chullo cap and woven poncho holding an open XO laptop on the cover, was the first such manual to be prepared in Spanish. Aside from demonstrating how to navigate basic functions of the XO, including mesh networking, recording video, and launching basic applications, the manual also suggested lesson plans built around local community-based customs and identity, including "Promoting Our Community Customs," "Describing Our Community's Principle Economic Activities," and "Recognizing Native Musical Instruments." It was that text, the first XO user's manual targeted for teachers written in any language, which began to earn Salas a degree of renown outside Puno. Since its initial publication in April 2009, it has been translated into English, French, and Arabic with digital copies downloadable from the Ministry of Education and the OLPC Foundation websites. Salas says, however, that she had little idea of the stir the book would create or the global attention it would garner: "The purpose was about getting material into the hands of other teachers. And the surprise was that there wasn't any other material on a global level like this. Which is why other [colleagues working in US universities] insisted that it be translated into English . . . . and I really was so pleased with that, because this was something that was done with teachers in mind, teachers from here, for Puno, for my people, but that turned out to be useful beyond."

Reflecting back on her willingness to take on the project, despite the requirement that it be completed in just a single month, Salas, who was trained as a statistical engineer, stressed her comfort in speaking to technologists and teachers alike. Underscoring what she sees as a need for greater convergence in skills and understanding between the communities of designers of education technologies and their key deployers, she said, "The large difficulty is that technicians are on one side, and teachers, professors are on the other. Where is there going to be an integration, a union? . . . That was why I put together the book. I know what a teacher needs and [could add] the technological part. . . . But what we do see, in many communities of free software, are technicians on one side and teachers on the other. This requires a lot of good communication [to resolve]."

She recalled becoming increasingly aware of the need for greater communication between such professional communities during educational conferences she attended that same year, when her emphasis on local deployments of the XO—and, in particular, its use among rural and indigenous communities—generated unexpectedly pointed interest her work. At a 2009 international conference for users and developers of the free software–based Squeak application, one of the dozen or so applications installed on XOs, she remembered being the lone participant who was able to address the issue of deployments with indigenous communities and multilingual, rural schools. And she recounts her surprise at realizing how few other participants had devoted work in similar contexts: "I was the only one that presented on students who had worked [with the XO] using native languages, children from rural Aymara communities. . . . So everyone in Brazil kept asking me how things were in Peru [with the XO]? Including representatives from [international governing institutions], who approached me and asked me how are thing going with this and that, and I realized there was really quite a lot of [general] interest from others to know how things worked with new technologies in Andean cultures. And I simply hadn't realized it before, since for us, it's part of our natural environment. But from abroad, it was different."

Salas speaks to me in a park courtyard outside of her school's gymnasium while a fellow teacher monitors her students' physical education class. She's dressed practically in a striped two-piece gym suit and she asks me to come and meet her students after she spends an hour or so talking to me about her work in Puno. Her dedication to the region and the work she accomplishes to bridging conversations between communities of educators and technologists—across the region and globally—is notable. She is among the few local teachers who routinely finds herself attending international conferences around new educational technologies. And it strikes me that given her international connections and the relative success her own work has now had among scholars and teachers working with such new pedagogical techniques abroad that she's likely had the opportunity to leave the region and move abroad.

Neither does she seem entirely satisfied with the work she has helped to coordinate in Puno, despite the larger enthusiasm it's helped to generate and the arguably global hype that's surrounded the OLPC project more generally. In speaking about the project's deployment in Puno, she stresses the work left to be done around building genuinely collaborative partnerships—and espousing the values-free software communities so frequently cited about nonhierarchy and openness—between the multiple parties invested in local deployments of the XO. Drawing together technology-centered cultures with indigenous cultures, national government offices with regional ones, transnational technology foundations and development-centered spaces with rural communities, and first-world academics with teachers from rural provinces, such projects bridge large differentials in questions of access to economic, technical, and social resources, power, and publics—topics that indeed rarely find themselves addressed. And, in more sobering tones, she discusses with some frustration how, too frequently, such questions of how to build collaborations and how to learn from and value the contributions of local (and especially rural) partners framed as peripheral generally have been seen as unnecessary to address or perhaps already adequately dealt with—rather than, as she sees it, only beginning to unfold.

Pointing to what she sees as the Ministry of Education's long-standing tendency to frame digital technologies as the "solution" to the question of educational reforms, she says, "The thing is that they thought, as they had [with other digital education initiatives] before, they would only have to buy the machines and it was going to be a *miracle*. And that was an error to see the OLPC as if in some spontaneous way learning would just emerge. If it emerges spontaneously, it happens through an instructor's encouragement and guidance." She says, however, that the same "error" in framing technological solutions was reinforced by the OLPC's network of technology developers, who similarly expressed disinterest in the role and work of local teachers and saw education as an activity that could be mediated with a well-designed technology alone: "Machines are only a complement, even [when they are] of a higher level. . . . Machines can't make changes themselves. Change has to occur through a teacher, and that was something I also said to [OLPC's representatives who came here]. . . . A teacher isn't about machines. A teacher is about what you make happen in the classroom."

She adds that despite such discouragement, she still works to advocate for other forms of local knowledge—in particular for the Aymara- and Quechua-speaking communities—as integratable into OLPC's pedagogical programs. "I tell them that we have Andean knowledge that grandparents have passed on and that functions, but it's not written. . . . And they tell me that you have to support this hypothesis, this theory on the Andean that you have with the recent scientific method, and then we can pay attention to you. The information already exists. But that's the barrier that comes from 'knowledge.' Unless we have large universities, if we don't publish papers all the time, and if it's not written in English, there's no respect."

Such frustrations, however, have not stopped her efforts around building local knowledge networks with teachers and local technology advocates in Puno or attempting to highlight the products of their work. Neither do such frustrations keep her from participating in global networks of OLPC advocates and developers. And she insists that such encounters across differences may also serve as learning opportunities for international technology developers and education reformers: "[They might] want everyone to work the same. But each culture assimilates the material in [its] own way. And you have to respect that, to build bridges, and not just say I am from the West and I know it all. You have to be able to know that 'you know too.' To be humble, [or at least] more humble . . . rather than presume these people are backwards or don't know anything. They do know, and know a lot. About the quality of life, and being human, even if it's a sense of life that's not [always] valued."

Back in Puno's municipal conference room where the software translation project is still underway, it is the strange mix of coca leaves and computers scattered on the table that calls one's visual attention, with local teachers from Puno, Quechua and Aymara elder and youth activists, and Escuelab technologists from Europe and the Americas collectively working its edges. Although the mood in the room is generally light, with volunteers clustering themselves into small discussion circles, and jokes tossed around over the strange and imprecise translation of terms among Spanish, Quechua, and Aymara, I can't help but think of the site as one of "awkward engagement." In Anna Tsing's (2004) framing of the term, it signals a space in which distinct cultural forces, social interests, and individual bodies meet in a "zone of cultural friction"—one in which consensus and the easy reply of givens are never taken for granted as means of erasing or dissolving differences—and "where words mean something different across the divide even as people speak" instead (xi). It meant to point to the friction simultaneously enacted along with the extension of "universal" appeals—such as, for instance, that around technology's universal distribution—that may draw and recruit diversifying interests into potentialized encounters but never completely dissolve the nonalignments, tensions, and debates associated with such universal-seeking acts to begin with.

Indeed, if global circuits of IT development more generally, and around the OLPC project specifically, have depended on the recruitment of partners to extend their project, they have sought to do so seamlessly, emphasizing the obvious self-evidence of the "answer" they offer. With the question they frame as one of "to do nothing" or "to partner and act" through technological distributions, the work of expanding partnerships—and acts of conversion with it—are meant to appear self-evident. Local actors invested in regional deployments of OLPC know, however, how thin such veneers can be. Attendant to the differentials of power, resources, and interpretations that persist across the network of collaboration, they insist on breaking from and interrupting the given operations of digital extension to speak through the various "unknowns" that still surround the deployment of global technology-based projects, such as OLPC's, asking the questions that, even despite the diverse interests drawn around such expansive and complex networks, remain conspicuously unvoiced.

There is, after all, no expressed assurance of IT's inevitable or self-evident drive toward progress in the despacho ritual Rodriguez and his fellow co-participants in the translation workshop collectively partake in. Neither, however, is there a call for broadening disconnection. It instead draws reference to contingency and uncertainty and the necessity of the guidance from the past in order to enter into a relationship with technological futures. Hours after the ceremony, Rodriguez stresses just this, telling me that as he sees it, technological change and its related complications are things his own community, and others like it across the region, will have to face: "Literally, we know it's going to enter . . . . It already is." He describes the connection made, however, between distinct temporalities and the need to draw them together to cultivate a space of mutual consideration in the face of future uncertainty: "As Andeans, for our communities . . . coca is a symbol of recognition of our ancestors. In order to begin a task . . . one has to thank the Pachamama [and so] . . . we always open activities with coca leaves that let us call upon spirits, thank them, and ask permission, so that the activities might turn out well."

The sentiment is one that is echoed by Aymar Ccopacatty Pike, a thirty-four-year-old US-born, Aymara visual artist and Escuelab resident who worked as one of the main coordinators of the translation workshop. Ccopacatty, who learned Aymara while returning back to Puno as a boy and spending summers at his grandmother's side, recalls having observed such ritual offerings since his youth. Describing the ceremony, he tells me, "It was about time and space, about asking for permissions . . . [and about] the essence, the spirit of ancestors in order to continue forth on the path. Because coca functions as a medium of communication of the present to the past, and for the Aymara, the future is in the past. It's behind us, which is a very circular concept—and the past is present."[6]

Such alternative framings highlight an attempt to rework the grammar and performative cultures that surround information technologies via operations of interrupt. They are notable precisely in their expressed effort to highlight the multilayered work that surrounds the stabilization of the "fact" of technology's universalizing expansion—and in their attempt to reveal it as one whose projected future of extension and progress is anything but obvious, natural, and inevitable. Or as Isabelle Stengers writes in describing the cosmopolitical challenge to modernity's universals, such moments of interrupt open to us the possibility of "reinvent[ing] the questions wherever we have been converted into believing in the power of answers" (2007, 80).

Notably, the alternatives posed are not ones that attempt to work against the notion of global connection or that pose the local and particular as antidotes to a problematically framed "universal." There is not, here, an attempt to reject new information technologies altogether or present them as the globalizing other to local peoples and native or minority cultures. Drawing on distinct material artifacts and explicitly working to create new cultural interfaces, they instead attempt to explore the possibility of developing other models of interconnectedness around IT, ones, for instance, that could attend to the particularities of history and context and extend a hand toward the past while still bridging, however tentatively, a dialogue with the future. It's here, in such newly forged zones, that the presumed givens and consensus of technology's universalizing future can begin to be unsettled and slowly, perhaps, give way to something altogether unexpected.

Rodrigo Arboleda, Chairman and CEO of OLPC Association, speaking on the means "to create true wealth in the twenty-first century" at the 2011 TEDxKids Conference in Brussels.

Conclusion: Digital Author Function

There are few signs that what allowed for digital universalism's rapid rise in contemporary Peru—or elsewhere, for that matter—will be waning anytime soon.

This is the thought that crosses my mind as I read the recent headline blinking across the top of my computer screen from a 2010 online edition of *Wired* magazine: "OLPC's [Nicholas] Negroponte Offers to Help India Realize $35 Tablet" (Paul 2010). The announcement, an unusual one by all accounts given that India had reversed plans several years ago to become OLPC's largest partner nation, came after the Indian government announced in 2009 that it would partner with its own national universities to develop a low-cost computer for use by students (BBC 2009a; Ribiero 2010). Now, photos of India's white-haired minister of communications and information technology, Kapil Sibal, proudly holding up a prototype of the seven-inch touchscreen tablet that he declared as "India's answer to MIT's US $100 computer" were circulating internationally. And with the international technology press abuzz about its expected release date before the end of 2011, it came surrounded with a significant degree of not only national anticipation but global excitement as well (*Economic Times* 2010; Katz 2010; *Times of India* 2011). And this, even before a single student user had even gotten to hold one of the first tablets.

The Indian venture, of course, is far from the only low-cost computer to enter the global education market in recent years. When MIT Media Lab founder Nicholas Negroponte first announced at the 2005 World Summit on the Information Society plans to begin development for a low-cost $100 laptop for use by elementary school–aged children worldwide, it was lauded as bold, groundbreaking, and inspirational for designing educational computing for "developing" contexts—and extending them to zones and contexts in which computers and high-tech hadn't previously been intended to go. In less than a decade, however, a rapidly growing number of similar commercial and nonprofit projects have been introduced to develop and extend low-cost, low-power computers for diverse global educational contexts. By 2008, just a year after OLPC's first official deployments began, more than two dozen such models had been launched (lilputing.com 2008). Intel's Classmate project, which distributed four million student-centered laptops at a cost of $400 to $500 a unit to partners in India, Portugal, Argentina, Venezuela, and Nigeria, entered as an early and highly visible competitor to OLPC (BBC 2008; Sniderman 2011). Today, however, several hundred low-cost, portable computers are available, including netbook models by Dell, Hewlett-Packard, Acer, and Asus (Kraemer, Dedrick, and Sharma 2009; *Washington Post* 2011). Such rapid growth around the development of low-cost educational technology models suitable for global markets—what had been considered a "previously neglected market" (James 2010, 382)—is an impact both fans and critics of OLPC credit it with (Charbonnier 2009; Gaines 2009). As development economist Jeffrey James (2010) has written in questioning the surge of national policies that privilege relatively expensive computing unit purchases over other potential educational investments, "For all the attention it has received in recent years, one might have thought that the idea of giving a laptop to each schoolchild in developing countries had been subject to intense scrutiny. Unfortunately, it has not" (387).

The "schoolchild from the developing nation" with mobile computers, indeed, has been the dominant globally circulating image around OLPC since the program began. Images of children in rural and remote zones, otherwise disconnected from the world and locked in spaces seemingly left behind by time but now, newly equipped with glowing, globally connected XOs, still (and all-too-familiarly) accompany press accounts and conference presentations for the OLPC. In a September 2012 missive published in MIT's *Technology Review*, Nicholas Negroponte highlighted the promising results of the latest experiments OLPC had undertaken—in a preview of a live presentation he would make at the 2012 Emergent Technology conference (Negroponte 2012; Talbot 2012). In between Negroponte's summary of OLPC's new deployments in Ethiopia, where boxes of XOs were simply left behind in remote rural sites, were images of children in dry, dusty villages or seated in straw-thatched huts, literally captivated and staring into their XO. Whether conveyed to audiences composed primarily of engineers, designers, policy makers, or a general public, such images are presumed to make evident the hopeful, inspiring (rather than troubling) power of engineering and universal appetites for technological solutions and design.

Rodrigo Arboleda, the Colombian-born chairman and CEO for the OLPC Foundation, similarly opens his presentations on the global "mission" of

OLPC, for instance, with a short three-minute video centered on Uruguay's program. The video cuts between clips of rural school children engaged in various everyday activities with their laptops—in class, walking on dirt roads, playing on farms and open pastures. In the last several years, Arboleda, along with key OLPC representatives including Walter Bender and Nicholas Negroponte himself, have been regular speakers in the international circuit of TED (Technology Entertainment Design) conferences that are renown for drawing crowds of tech-savvy, global professionals, planners, and entrepreneurs. Taking to the stage at the 2011 TED conference in Brussels, Arboleda followed his short video segment on the Uruguayan Program with a reminder to his audience of OLPC's commitment to universal inclusion: "The challenge is to not leave people out—and to have every child of primary school age in third world countries have the same opportunities [and] access to knowledge in [the same] quantity and quality as the most privileged child in New York, Tokyo, Berlin, or Brussels. That is the challenge!" Arboleda had made similar pronouncements at earlier TED conferences in Rio de Janeiro and Miami. Pacing atop a high stage, with images of South American students toting their XOs through a rural countryside still flashing behind him, Arboleda warned his cosmopolitan audience in Brussels that the fate of "third world" economies would be affected by decisions governments made today around educational technologies. Asserting that the essential problem of the Global South was their "insufficient" production of "engineers, innovators, and scientists" and the valuable forms of IP knowledge work generates, he added, "In countries in Latin America and Africa, we are still in love with [material] commodities—and we think that producing commodities is going to bring us into the twenty-first century. Which is not the case." Pausing to emphasize what he framed as the self-evident global solution to twenty-first-century commodity production, he added, "We need to create wealth for the twenty-first century, and it has a specific name—which is Intellectual Property."

There are, of course, more than a few strategic omissions in Arboleda's presentation. For all his invocations of science, technology, and their roles in innovating twenty-first-century commodities, he makes no mention of the new state initiatives that are already targeting rural producers in the global south as potentially new producers of IP titles and IP-able goods discussed in the first section of this book. Just as he makes no mention of how such accelerated IP titlings have spurred controversy in Peru, or generated critiques among local producer populations like ceramicists from Chulucanas. And perhaps most tellingly, he makes no reference to the tacit Western-centricism in his framings of "innovation" and "science," their projected absence in the non-West, and the assumption that cultivating more innovators and authors of IP—as it has in the West—would be the essential step to "creating twenty-first-century wealth" in the Global South.

Neither does he mention the particular relationships OLPC relied on for their local deployments across Latin America and around the globe. He makes no reference to the project's dependencies on the extensive contributions and expertise of globally dispersed, local partners—including the regional FLOSS networks and situated educator and indigenous networks discussed in section two of this book. And he makes no reference either to the distinct modes of collaboration expressed through new cultural interfaces fostered through such networks—that bridge indigenous tradition with the high-tech and local interests with the global—that push forth other visions for global connection and technological futures.

He is, however, clear that his prescriptions around digital and information-based properties could be leveraged as not merely evolved strategies for national economic and technological development but for individually targeting reforms as well. "Something that is already [developed] in the first world is a language to transmit information and transmit knowledge that has been around for more than 50 years, but that we have only started to scratch the surface of in Latin America and Africa . . . which is the 0 and the 1. . . . [T]his is the new alphabet, and this is what we are trying to format. . . . Because in the Digital Age, a child in Africa has the same opportunities as a child in Brussels if we provide him with the right tools."

Arboleda, of course, does not merely project an image of the "third world" as a space in need of intervention and aid during his presentation. He at once delivers an image of the "first world" back to his audience—one that reflects it as a space of ready and able global problem solvers, tech-savvy innovators, and knowledge-embracing experts less invested in securing international dominance through force or coercion (as may once have been the case) than in acting on behalf of a greater good and the general progress of humankind. And if there is once again a claim made for the right to intervene in "the third world" using modernity's tools, it is not so much with the explicit aim to globally "civilize" and "enlighten" as to ensure the optimal generation of individual creativity and reformed human potential.

Cast in the role of global problem solver, this idealized twenty-first-century actor may bear similarities to those of earlier Western universals—animated, for instance, through modernity's civil, gentlemanly, Kantian cosmopolitans or enlightened, rational men of science. The objective here, however, is no longer to transform the nation state or physical world into modern ideals as much as to remake human cognition itself. As Arboleda explains the new potentials to optimize the mind as an information processor and to cultivate new inventors and innovators through the shift from analog to digital modes of cognition: "We are still pegged to the 26 symbols of the alphabet to transmit information and knowledge. We have forgotten that there is new mode of transmitting information that is only 2 symbols, the 0 and the 1. . . . What we need to do is to create a different profile of a brain. *A brain that is geared towards innovation, exploration, invention.* And that only can be done if you empower the child from the very beginning [with technologies] for children to explore. So the game here is to *create inventors and innovators.*"

Indeed, OLPC has spurred new interest in cognitive sciences around the mapping of the neural processes in childhood education. OLPC's chief education officer, Antonio Battro, an Argentine-born medical physician, has been particularly active on this front. In his article "The Teaching Brain," published in the journal *Brain, Mind, and Education,* Battro (2010) argued that "teaching" should be recognized as a fundamental characteristic of the human species. As such, it should be understood less as a professional skill set reliant on specialized training or institutions of higher learning—and should be recognized instead as an innate capacity, already present in children, that likely enabled humanity's survival as a species. (Based on such thinking, he even posited that this should warrant "our species [be called] *Homo sapiens docens*" [29], docens meaning "teaching" in Latin.)

Such work has been critiqued for presuming the "psychic unity of mankind" (Smith 2011, 18) and for promoting a vision of a "teacherless world" filled with laptop-supplied, self-teaching children. Battro, however, has remained steadfast in his defense of the field and its blend of neuroscience, education, and psychology. Citing research undertaken in collaboration with such institutions as Harvard University's Mind, Brain, and Education program (Battro, Fischer, and Pierre 2008), the University of Tokyo and Hitachi Labs (Aoki, Funane, and Koizumi 2010), and the International Mind, Brain and Education Society, he's dismissed attempts to question the legitimacy of the field asserting, "'Neuroeducation' is now *a fact* and laptops are a mobile and portable laboratory for recording many brain activities. We are entering a new digital era, with many more intellectual and technical tools than those provided by the teaching and learning restricted to a classroom or computer lab" (Battro 2011).

Battro is not alone in such proclamations. A growing number of publications that target a general audience with the promise of being able to help them optimize cognitive functioning today regularly fill the shelves of bookstores and best-seller lists. Academic neuroscientists and cognitive specialists have begun to target general audiences with texts, such as

UCLA neurobiologist Dean Buonomano's 2011 *Brain Bugs*, that offer recommendations for improving brain functionality. Buonomano (2011), for instance, informs readers in the opening pages of his text that "the brain is an incomprehensibly complex biological computer . . . designed to acquire data from the external world through our sensory organs; to analyze, store, process this information and generate outputs—actions and behaviors— that optimize our chances for survival and reproduction. But as with any other computational device, the brain has bugs and limitations" (2). Such claims are key to digital universalism's appeal. It turns in good part, after all, on its promise to perfect something as seemingly ubiquitous and individually accessible as information processing itself. Unlike prior Western universals, it could undertake its reforms without the same dependence on (or burden of) institutions and authorities associated with prior universal projects. Thus, initiatives such as OLPC can emerge as a stand-in for global reform via an unfettered individualism, freed now from the inefficiencies of established institutions of religion, states, markets, or even science.

Arboleda fashions one final example of this in closing his presentation in Brussels. As he ends his talk, he flashes images of a number of OLPC-partnering classrooms before his audience: one of a classroom of veiled girls from a nonidentified nation, each supplied with her own XO, another of school children in Colombia's "guerilla-infested" mountains, another of rural school children in Peru. Below the image of the Peruvian school children triumphantly appears the text, "The end of isolation." He avoids any mention of the particularities of history or context of any of the locales he displays on the screen. He doesn't mention how the schools he references in Peru, for instance, had recently seen some of the largest teachers' strikes in decades—where schools were shut down for months, town roads were blocked, and protests called for the entry of security forces as flows of goods and medical supplies stopped outside town borders (Associated Press 2007; Collyns 2007). No mention is made either of the newly implemented law that had sparked demonstrations in 2007 and that adjusted teachers' salaries according to their students' scores on national exams. And no mention is made of the mutual distrust that state officials and rural communities historically held for each other and that had quietly been simmering since the state's armed conflict with the Shining Path's rurally organized rebellion officially ended in the 1990s. It was as if the sites and communities partnering with OLPC hadn't had any significant existence (in digital or otherwise) until its collaborations with OLPC began. Or as if they had been disconnected from the world and historical memory until the XOs had arrived. And now that they had, audiences—interlinked, online, and attentive to the message of networking's global potential—seemed unable to get enough of it.

Little wonder why, however. The story of digital innovation and authorship as practices that could liberate the creative potential of individuals and at last enable the unrestrained, true development of the self has proven after all a hard one to resist. To audiences connected online around the world, able to swap screens and shift artfully between acts of digital work and play, this was as much their story as it was that of remote rural school children. And it was a story proving ever harder to turn away from. Indeed, against a backdrop of flaring tempers and pitched suspicions, digital authorship, innovation, and the notion of optimizable cognition could emerge as at least one safe territory where all interests could miraculously converge and lean toward a glowing future in formation.

It's helpful to recall that what's brought to mind in the notion of authorship—as the dedicated state of producing creative, original content—did not always manifest in its contemporary form. What is recognized today in authorship as the labor of individual genius or the expression and invention of an "original" work that hadn't existed before is itself a product of various historical processes. Michel Foucault (1977) underscored this social contingency in positing the notion of an "author function," writing that although what may be recognized as the "author" has changed over time, the function of authors to "characterize the existence, circulation, and operation of certain discourses in society" remains remarkably consistent (124). What is natural and inalienable to authorship, Foucault insisted, is its existence in proximity and relation to centers of power. As such, it is a function that always reveals more about who or what holds the power to "authorize," than who the author actually "is." Rather than attempt to settle questions of "who is the real author" or is there "proof of his authenticity?" then, Foucault stressed that the unspoken conundrum remained: "What matter who's speaking?"

Said otherwise, we might ask: what matter *to us* that the author speak? Why, that is, do audiences persist "in granting a primordial status to writing . . . to reinscribe in transcendental terms the theological affirmation of its sacred origin or a critical belief in its creative nature" (Foucault 1977, 120)? If there is a diagnosis Foucault issued in the query, it is not on authors or even their authorizers, but on the public's own relation to authorial status—and its curious enabling of the author function's persistence. Just as it may reveal more about who or what holds the power to authorize content than about who it is that purportedly created an authored text, so, too, might the author function oddly reveal more about audiences and the author-seeking public than about authors themselves.

And today, perhaps more than ever, the persistence of a global author function urges explanation. It is in the contemporary moment, after all, that we hear of authorship's entry into a new age of pointed crisis, when new modes of digital expression and networked exchange threaten the former national, legal, and moral conventions that had stabilized authorship's integrity. In a network-ready world, where unlicensed duplications, pirate acts, and the modifications of "authentic" forms were as simple as the pressing of a computer key, the privilege and protection once promised authors, as actors endowed with the "origin" of creative acts, have come undone. And it is curiously at a moment when traditional authorship appears more threatened than ever that a fetish for new digital authors and innovators seems to have at once become instantly pervasive.

The author function, however, seems to have always rested on an odd contradiction: being as much about the dream of coming into power's graces as it has been about being able to create exceptional members of society able to make statements in its name. It has been, that is, as much about the collective hope of being named and recognized as worthy and exceptional as about exercising the power to distinguish and mark worth. And just as not all articulable statements are worthy of having authors, neither are all beings worthy of being acknowledged as articulators of "authorizable" content.

In the modern era, the author function's ability to assign recognition and privilege to authors demonstrated not only that enlightened creative acts of civil discourse were ideal practices for modern subjects but also that it was possible to assess discursive acts—to objectively distinguish "right" ones from "poor" or "wrong" ones. It remained, however, a sovereign power that acted as the primary custodian and verifier of true authorship. And it remained, too, largely men of privilege who were extended opportunities to even be considered as possible authors.

The logics developed via the global information economy and the growing diffusion of networked technologies—as the extension of IP experiments like those in rural sites like Chulucanas discussed in the first section of this book—could turn such hallowed traditions on their heads. If digital authorship could declare itself an improved version of traditional authorship, it could do so in part because it no longer relied on established powers to authorize, stabilize, or protect authorial status. Promising to dispense authorship's conventional gatekeepers and unravel the limiting categories by which privilege had once been determined, the networked age proclaimed the liberation of unbounded individualism and democratic, creative opportunity, and in evermore diverse sites. In Chulucanas, IP promoters could now declare that the possibility of authorial conversion lay in the hands of each individual artisan and their own creative enterprise. Given that creativity and intellect—the raw materials of IP-centered production—were universal human attributes, innovation with IP could now be framed as broadly inclusive and embracing of the voices of new classes of collectively potentialized authors. IP-centered innovation, that is, could pool all actors' capacities together, enabling the rural poor to participate in markets alongside the urban rich, and allowing the formerly marginalized to compete alongside the historically privileged.

The message of IP promoters in sites like Chulucanas remind us that there was at least one further liberatory claim networked authorship was said to enable: that all subjects—rural and urban, brown and white, poor and privileged—could now be recognized for new creative capacities and as agents responsible for their own uplift would crucially free states of such responsibilities. The lesson they shared being that human capacity, once placed on a democratically equalized playing field, could economize the expenditures of the state. Rather than being responsible for creating common architectures of civic entry and articulation, states could now spare their resources. In the name of maintaining incentives for social advancement, in fact, they could be obligated to reserve investments for only those particular subjects who did distinguish themselves from the common pool as exceptional authors, inventors, and competitors.

The prospect of convergent interests and eternal progress here, however, is not one that obliterates the conditions of war or conflict. This too, was a lesson new inter-community tensions that arose in the wake of the state's IP-centered investments in Chulucanas imparted. Civil strife found fresh ground and stakes, waged now between classes of actors who posed new forms of competition to one another as supposed inhibitors to another's progress and advancement. What emerged as the defining symptom of such IP-based investments were articulations of newly produced struggle— held now between actors who once cohered under the categories of society, community, neighborhood, family (or even the self). And such internal battles within, of self against self—and done in the name of reform—have never appeared so productive, intensively spectacular, or—as the accounts of artisans in Chulucanas attested to—so pervasively shared.

There are, still, other futures possible around global digital connection. The critical stances and interventions developed around educational technologies like OLPC deployed across the provinces, like the emergent collaborations newly fostered between FLOSS advocates and coders, rural teachers, and indigenous language activists around FLOSS projects—that were the focus of the second part of this book—represent efforts that push for other forms of digital connection. And while such citizen-based efforts express an openness to experimentations in the present—not unlike new initiatives undertaken by the state or global projects like OLPC—their engagements instead are ones that urge other imagined futures that still insist upon a linking (and indeed, often uneasy linking) to the past.

This is something the thirty-four-year-old Aymara artist Aymar Ccopacatty reminds me of as we sit in the lobby of a hostel in Puno where several of the Escuelab participants have spent the past week while volunteering for the Sugar Camp activities. Although most of the young technologists and artists who arrived in Puno for the conference are now preparing to travel back to their cities of residence—La Paz and Cochabamba, Bolivia; Barcelona, Spain; and Cuzco and Lima, Peru—Ccopacatty is planning an extended stay in Puno. Ccopacatty had been one of the organizers of the Sugar Camp sessions to translate the XO's free software interface into the Andean languages of Quechua and Aymara. He explains that he typically splits his time between Peru, his father's home country, and the United States, his mother's and his own nation of birth—and tells me he expects his next several weeks to be spent in his father's home village, Acora, where family land is still maintained and where his father helped his farm before gaining renown as a visual artist and sculptor in his adult life. It was in the Acora home where the younger Ccopacatty first learned the traditional weaving techniques he uses in his own artwork today and where—even before he learned Spanish—he learned to speak Aymara from his grandmother. He tells me that she herself never spoke Spanish and taught him her native language during his summer vacations. It was his grandmother, too, he continues, who taught him to plant, to grow potatoes and other crops, and "live off the land." And although she passed away several years ago, he still speaks of her in the present tense.

Explaining his plans to extend his time in Puno, he tells me, "I know that I have a grandmother that doesn't speak Spanish, who speaks Aymara, and who knows how to weave and to sew, and is an agriculturalist, who lives from the land in this way. . . . I know that this world is going to change very quickly, and when I compare this balance with [what's to come], staying is not even a question."

Backpacks and equipment cases of Escuelab's "activistas ambulantes"— traveling free software activists—accrue against the walls beside us as I ask Ccopacatty about his expectations for the translation project and for the OLPC-centered collaborations that had unexpectedly drawn some four hundred teachers into Puno for the past week.

Rather than give me a direct answer, however, he settles into another story, this one that another Acora elder had passed down to him, of a traveling man from Europe who had found his way to one of the local villages of Puno's high plains. Arriving, the visitor was surprised to find a hunger-stricken Andean community, unable to farm the land because of the bitterly cold climate of the altiplano, that was only able to live from the slaughter of their livestock. "And so this man spoke to the community and said 'I will bring you cattle, European cattle of the first class' . . . and indeed, two years later, some very special cattle did arrive, of pure blood, something truly marvelous. But then the condors arrived, and the largest condors began to steal away the smallest calves. And the town became furious and said, 'we can't let something like this happen to something so important. . . . And so they took a dead sheep and filled it with poison, and threw it into the air. And sometime later, the town's annual festival arrived—a time of year when condors come down from the heights to altitudes closer to those of the towns, to find a mate and continue the cycle of life. But the date came and went, and the condors didn't come down, and the people became frightened. So they decided to climb to see what had happened to the condors, and they saw that the condors had died, and their bodies were strewn everywhere because of the poison." He pauses then to consider his tale. "And so I suppose this is my way of saying that sometimes there are good intentions—the man who brought the cattle had truly good intentions—but the unintended effect was a massive massacre of condors."

Addressing digital technologies more directly, he goes on to tell me that he still believes that "there are cases when the machine stops being a colonizer and can enter into another space, to become something more democratic. . . . When it enables at once practices that re-create culture so that culture may be transmitted. With the XO you can learn to read and write in Aymara and Quechua. But this has to [be done while] recognizing that it has the same source of life. It has to be able to follow, even [as it moves ahead], and that can join together all the assorted little parts that compose it. . . . It can't attempt to occupy the space ahead and beyond, because that would be to once again kill culture, rather than support it. So, it is a thread so thin between the act of killing and that of support with technology. It's something that there is no clear black-and-white answer to. Everything is gray."

Listening to Ccopacatty speak, I can't help but recall his rumination from days earlier during the translation session with indigenous elders and young hackers, an event discussed in chapter 6. Over a table scattered with coca leaves and computers—both media for speaking with outside forces— he mused of the Aymara vision of temporality as one where "the future is in the past." Here, as then, he didn't balk at the notion of encounter, even across radical spaces of global difference. Neither did he attempt to frame local spaces as zones of purity that offered a means of escaping global forces. Similar to other critics of dominant forms of universals before—from postcolonial scholars who decried the effects of the West's cultural universals (Escobar 2007; Mignolo 2007; Wallerstein 2006) and from science studies scholars who question science's claim to natural universals (de Castro 2004; Latour 2004b; Law 2007; Stengers 2007, 2011; Verran 2007, 2011)—his was not a retreat to a sanctuary of relativism. Such an attempt to escape the space of universals through the defense through claims to irremediable, incommensurable difference would insist that no form of mutuality could ever be achieved. Here, however, was an explicit call for not only encounter and the global minglings or confrontations they would entail but also for dialogue and mutual learning in the process of such indeterminate meetings.

His words linger with me when I learn some weeks later that he joined regional protests that broke out in Puno, where more than two hundred communities and thousands of area residents took to the streets to demonstrate against new transnational mining concessions in the area. His caution, like those of the diverse citizens actors discussed in part II of this book, sought to develop a language of political reform and critique via engagements around digital networks, FLOSS, or educational technologies. Their efforts emerged as a call for a kind of pluriverse that reads "socionatural" encounters—particularly those drawn together via digital networks—as always political ones (de la Cadena 2010) that recognize an obligation to account for and engage with multiple past histories that are drawn together in such meetings, however messy such encounters might be, before making any claims for the future. And even while confronting urgent attempts by dominant digital classes to shake off the burden of the past in a rush toward the future, their efforts echoed the challenge posed by subaltern theorists to work towards the "building of other possible worlds" (Mignolo, 2007, 479) precisely through a constant reconnection to the past. Or as Walter Mignolo had put it before: to "connect pluriversality [as] different colonial histories entangled with imperial modernity into a universal project of delinking from modern rationality and building other possible worlds" (497).

The networked citizen collaborations discussed in the second half of the book offer reminders too of how cultivating interconnections between worlds can aid processes of knowledge production. Their activities and investments underscore arguments made by critics of scientific universals, including Isabelle Stengers (2011), of experiments and the construction of "experimental reality" as always acts too of "creat[ing] rapports" and new modes of relating (Stengers 2011, 52). Such a framing would entail an acknowledgement of the plurality of sciences and a pluralization of modes of concern that surround processes of knowledge production. It would, in other words, not only mean that actors invested in knowledge and technical production acknowledge the plurality of epistemologies and forms of knowledge production but also to see such work as undeniably and valuably invested in the production of relations across social, natural, and technical bodies. This, even if dominant forms of digital universalism—enthusiastically oriented as they are towards the ecstatic promise of the future—might promise new forms of social engagement that could be experienced less as unpredictable or undefined forms of relating than ones that could enable an easy escape from the messy complications of unknown encounter—or indeed, encounter with a past one would rather forget.

Their investments remind us too that developing new modes of interrelating and techno-civic forms of rapport could not come sooner. For whatever dream ICT promoters within the global digerati were selling, whatever came packaged in the glossy green and white plastic cases of OLPC's XOs that had captured not just the Peruvian state's attention but also that of networked audiences around the world, it would surely not be the last machine to do so. Just as with the deployment of IP titles in Chulucanas or that of other information-based technologies and networked resources for development before it, what is certain is that there will be another technology of reform to follow. Another liberation machine that shall be willed to captivate our collective imagination and hold us in the promise of its reflective thrall.

Here, then, care must be taken in enlivening local relations—to engage, that is, with considered treatment of the past and enroll local histories. Such work could be undertaken in an effort to build new relations that avoid any goals of assimilating difference in order for such others to be recognized as part of a digital contemporary. This would entail an active acknowledgement of past ruptured ties, broken trusts, and unresolved conflicts that still shape the contexts of relating. And it would entail a continual reposing of questions in the face of such acknowledged fragmentations: how might we foster engagements to ensure that universal forms be "charged and changed by their travels" (Tsing 2004, 8) across spaces of pronounced difference? And how, too, might the aims of consensus so common under universal flows give way to something else—to an enabling of the kinds of rapports and relational spaces to allow a mutual "learning from" (Stengers 2011, 63) so that diverse parties can emerge instead?

It should be clear, then, that this project does not mean to foreclose networked connection as an inevitably colonizing force, despite the haunted record that surrounds universal assertions. I share the aspirations of other critics of modernity's universalistic thrust that a more conscientious, humbled version of a gesture reaching toward "global connection" may be recovered for our networked age. Historian Dipesh Chakrabarty (2000) emphasized such a modest prospect in framing the international spirit of socialism, postcolonialism, and Marxism as examples of "engaging universals." Even though such bodies of knowledge aim toward global extension, they operate to decenter the totalizing powers of Western universals precisely in their ability to enable locally situated encounters and mutations as they travel. Writing of this knowledge's unique capacity to speak across diverse spaces and the means by which it can productively be called on to represent the interests of the elite and nonelite alike, he notes the "universal" less as the guarantee of uniform replication and instead as "a highly unstable figure—a necessary placeholder in local attempts to think through and re-engage questions of modernity" (xiii). Likewise, ICTs and digital networks, in their own universal movement through space, might be said to offer means for diverse populations to distinctly think through the multiplying questions of futuricity, unstable though such projections may be, and filled at once with pitched aspirations and unquiet anxieties.

Notes

Introduction

1. See WSIS newsroom at http://www.itu.int/wsis/geneva/newsroom/index.html and http://www.itu.int/wsis/tunis/newsroom/index.html.

2. WSIS would further explain the shift in its General Assembly Resolution as one that sought to include "the active participation of all relevant United Nations bodies . . . and encouraged other intergovernmental organizations, including international and regional institutions, non-governmental organizations, civil society and the private sector to contribute to, and actively participate in, the intergovernmental preparatory process of the Summit and the Summit itself." http://www.itu.int/wsis/basic/multistakeholder.html.

Chapter 1

1. Sendero's entry into the northern provinces of Peru, in the Piura and Cajamarca regions, was actually minimal, particularly compared to the violence communities in Peru's central and southern Andes and jungle experienced. Much of this was due to the work of peasant-organized "police" groups, ronderos campesinos, in Peru's northern provinces. So effective were they in deterring Sendero's entry that they eventually received sponsorship from the Peruvian government to continue this work. What's interesting, however, is that despite the minimal impact of terrorism in the northern provinces, it's nonetheless summoned to justify the intensive intervention of state and market forces there.

2. Literally translated, "I worked like a red ant." Its translation in English is clunky and awkward but in Spanish it is lovely. It captures the sense not only of hard, physically demanding labor but also the sense of determination, commitment, and duty on the part of the worker. Labor here is not mindless, not automatic, but is an expression instead of the worker's commitment to a larger community whose purpose is served by these individual acts of labor.

Chapter 4

Sir Edgar Villanueva Nunez
Congressperson of the Republic

Dear Sirs:

First, we thank you for the opportunity to inform you about how we have been working in our country to benefit the public sector, always seeking the best alternatives to realize the implementation of programs that allow the consolidation of modernization and transparency initiatives by the State. Specifically, based on our meeting today you learn of our international advances in the design of new services for the citizen, beneath the sign of the model State that respects and protects the rights of the author.

This action, as we discussed, is part of a worldwide initiative and today there are diverse experiences that have allowed collaboration with programs to help the State and the community to adopt technology as a strategic element to impact the quality of life of the citizens.

On the other hand, as we left things in our meeting, we joined the Forum that took place in the Congress of the Republic on the 6th of March, based on the proposed law that you lead, where we could hear the different presentations that today bring us to present our position so that you have a broader panorama of the actual situation:

1. The project establishes the obligations that every public organ should exclusively use free software, which is to say open source, which transgresses the principles of equality before the law, nondiscrimination, and the rights to free private initiatives, liberty of industry and of contracts, protected by the Constitution.
2. The project, by making the use of open source software obligatory, establishes a discriminatory and noncompetitive treatment in contracting and the acquisitions by public bodies, contravening the principles at the foundation of Law 26850 of Contracts and Acquisitions by the State.

Chapter 5

1. PLUG's website: http://www.linux.org.pe.

2. APESOL's website: http://www.apesol.org.

3. Such theory of play practices also draw from psychoanalyst David Winnicott, who has drawn attention to the latent power behind play spaces. Defining "play spaces" as a separate realm between those of the individual's internal psychic experience and the external social world, Winnicott conceived of play spaces as a transi-

tional ground between an individual's subjective reality and the external, objective world. There, the individual would have the freedom to engage in a kind of creative "reality testing" whose result could be to enhance the ability to "recognize and accept reality" (Winnicott 1971, 3). More than existing as a space that allowed an individual to come to terms with, and adjust the self to, a "real" social world, however, Winnicott suggested that play itself was intermeshed within circuits of potentiality and creativity that could serve more purposes than aiding the uncoached psyche to cope with a distinct reality. Kindling such unrestrained energy, play functioned as a practice that only barely managed to contain and direct itself toward a synchronization between and assimilation of the subjective and external worlds. Or as he writes, "Play is immensely exciting. It is exciting not primarily because the instincts are involved, be it understood! The thing about playing is always the precariousness about the interplay of personal psychic reality and experience of control of actual objects. This is the precariousness of magic itself, magic that arises in intimacy, in a relationship that is found to be reliable" (Winnicott 1971, 47). Play, by his formulation, operates through the unleashing of an indeterminate, dynamic body of forces that may be channeled for purposes of individual, self-adaptation, but only with a considerable degree of uncertainty.

4. SQL stands for "structured query language," a popular programming language used to build databases.

5. RAD stands for "rapid application development," which refers to a set of software applications and development methods typically used to build software prototypes and to deliver those prototypes under fast deadlines. The use of RAD tools is typically thought of as a practice oriented to generating corporate solutions.

6. PHP originally stood for "personal home page" and is a general-purpose, free software scripting language widely used for Web development. It is the free software alternative to Microsoft's ASP language.

7. ASP stands for "active server pages." It is a Microsoft technology for displaying dynamic web pages.

8. RPM stands for "Red Hat package manager" and is a storage file for Linux programs.

9. CVS stands for "concurrent versions system," a system that allows you to archive old source code and log changes.

10. Accessed September 1, 2005, from http://themes.freshmeat.net/projects/wmbrushedturquoise.

11. A compiler is a software program that converts a coder's source code into digital binary language that a computer can read. Once compiled, code is less easily "readable" by its coder and appears even less intelligible, as Ognio mentions, to noncoders such as his parents.

12. One of the most popular licenses for free software applications.

13. In fall 2004, Gutierrez and other students who had advocated free software use inside the Universidad de Lima founded the Study Circle ULIX (Grupo de Usarios de Linux/BSD/UNIX de la Universidad de Lima, http://aurealsys.com/~ulix). The group's recent activities include building a network or cluster of Linux computers for university use, meetings with the university's computer science and engineering professors on supporting and incorporating Linux use, and planning a university-centered conference on free software with members of local Linux user groups, including PLUG.

Chapter 6

1. See the website and wiki for Escuelab: escuelab.org and http://wiki.laptop.org/go/EscueLab.

2. See the website for the Ministry of Education's OLPC project: http://www.perueduca.edu.pe/olpc/OLPC_Home.html.

3. See the website for the "2nd International Conference—Sugar Camp Puno 2011—Amtawi Digital": http://escuelab.org/category/tags-etiquetas/puno.

4. See EduJam's English-language website: http://ceibaljam.org/edujam2011_en.

5. Arboleda's audience for this particular talk is indeed, many times larger. The Rio talk itself was reported to have gone viral online and its popularity has since made him a routine speaker on the international TEDTalk circuit, with the same talk having been given to live audiences in Miami and Brussels.

6. Aymara framings of time have received new attention from cognitive scientists and linguists as well. For more on this, see Nunez, Rafael, and Eve Sweetser. 2006. "With the Future behind Them: Convergent Evidence from Aymara Language and Gesture in the Crosslinguistic Comparison of Spatial Constructions of Time." *Cognitive Science* 30: 401–450.

References

Abbate, Janet. 2000. *Inventing the Internet.* Cambridge, MA: MIT Press.

Abrahamson, Mark. 2004. *Global Cities.* Oxford: Oxford University Press.

Adas, Michael. 1989. *Machines as the Measure of Men: Science, Technology, and Ideologies of Western Dominance.* Ithaca, NY: Cornell University Press.

Ahtisaari, Martti. 2011. Preface. In *Peacebuilding in the Information Age: Sifting Hype from Reality,* ed. ICT4Peace, The Berkman Center for Internet and Society, and The Georgia Institute of Technology. http://ict4peace.org/wp-content/uploads/2011/01/Peacebuilding-in-the-Information-Age-Sifting-Hype-from-Reality.pdf.

Alarcon, Daniel. 2008. Life among the Pirates. *Granta,* 109. http://www.granta.com/Archive/Granta-109-Work/Life-Among-the-Pirates/1.

Alonso, Ana Maria. 2005. Territorializing the Nation and "Integrating the Indian": "Mestizaje" in Mexican Official Discourse and Public Culture. In *Sovereign Bodies: Citizens, Migrants, and States in the Postcolonial World,* ed. Thomas Blom Hansen and Finn Stepputat. Princeton, NJ: Princeton University Press.

Alvarez, Marco, Baiocchi, Jose, and Pow Sang, Jose Antonio. 2008. Computing and Higher Education in Peru. *Inroads—SIGSCE Bulletin* 40 (2): 35–39.

Alvarez, Sonia, Evelina Dagnino, and Arturo Escobar. 1998. *Cultures of Politics, Politics of Cultures: Re-Visioning Latin American Social Movements.* Boulder, CO: Westview Press.

Anan, Kofi. 2005. Preface. In *Information and Communication Technology for Peace: The Role of ICT in Preventing, Responding to and Recovering from Conflict,* ed. Daniel Stauffacher, William Drake, Paul Currion, and Julia Steinberger. New York: United Nations Information and Communication Technologies Task Force.

Andean Currents. 2007. Peruvian Teachers Protest Turns Violent, July 13. http://www.andeancurrents.com/2007/07/peruvian-teachers-protest-turns-violent.html.

Andersen, Lars Bo. 2011. A Travelogue of a Hundred Laptops. Master's thesis, Aarhus University.

Anderson, Benedict. 1983. *Imagined Communities: Reflections on the Origins and Spread of Nationalism*. New York: Verso.

Anderson, Chris. 2006. *The Long Tail: Why the Future of Business Is Selling Less of More*. New York: Hyperion Press.

Anderson, Chris. 2010. *Free: How Today's Smartest Businesses Profit by Giving Something for Nothing*. New York: Hyperion Press.

Andina. 2010. Exportaciones en abril crecieron 44% y ascendieron a US$ 2,575 millones. Andina.com, June 3. http://www.andina.com.pe/Espanol/Noticia.aspx?id=CFdxZn6N0vQ=.

Aoki, Ryuta, Tsukasa Funane, and Hideaki Koizumi. 2010. Brain Science of Ethics: Present Status and the Future. *Mind, Brain, and Education* 4 (4): 188–195.

Appadurai, Arjun, ed. 1996. *Modernity at Large: The Cultural Dimensions of Globalization*. Durham, NC: Duke University Press.

Appadurai, Arjun, ed. 2001. *Globalization*. Durham, NC: Duke University Press.

Arbodela, Rodrigo. 2011. Children: A Mission, Not a Market. Tedx Rio Conference Presentation, February. http://riogringa.typepad.com/my_weblog/2011/03/tedx-brazil.html.

Associated Press. 2007. Peru Teachers Strike to Protest Competency Testing, July 5. http://www.iht.com/articles/ap/2007/07/06/america/LA-GEN-Peru-Protests.php.

Atkinson, Rowland, and Gary Bridge. 2005. *Gentrification in a Global Context: The New Urban Colonialism*. New York: Routledge.

Auletta, Ken. 2010. *Googled: The End of Work as We Know It*. New York: Penguin Press.

Bajak, Frank. 2007. Laptop Project Benefits Peruvian Village Children. *USA Today*, December 26. http://www.usatoday.com/tech/products/gear/computing/2007-12-26-one-laptop-peru_N.htm.

Bakhtin, M. M. 1984. *Rabelais and His World*. Bloomington: Indiana University Press.

Bakk-Hansen, Heidi. 2008. PERU: XO Machines Grant Users a Different Sort of Life. OWLInstitute.org, January 2. http://www.owli.org/oer/node/2458.

Barry, Andrew, Thomas Osborne, and Nikolas Rose. 1996. *Foucault & Political Reason: Liberalism, Neoliberalism, and Rationalities of Government*. Chicago: University of Chicago Press.

Battro, Antonio. 2010. The Teaching Brain. *Mind, Brain, and Education* 4 (1): 28–35.

Battro, Antonio. 2011. Comments on Jeffrey James' OLPC Critique. Blog.laptop.org, May 8. http://blog.laptop.org/2011/05/08/comments-on-james-critique.

Battro, Antonio, Kurt Fischer, and Léna Pierre. 2008. *The Educated Brain. Essays in Neuroeducation.* Cambridge, UK: Cambridge University Press.

BBC. 2008. Venezuela Splashes out on Laptops. BBC.com, September 30. http://news.bbc.co.uk/2/hi/technology/7642985.stm.

BBC. 2009a. India Plans Cheap Laptop Option. BBC.com, February 3. http://news.bbc.co.uk/2/hi/7864806.stm.

BBC. 2009b. Internet Brings Events in Iran to Life. BBC.com, June 15. http://news.bbc.co.uk/2/hi/middle_east/8099579.stm.

BBC. 2011. Peru: Indigenous Protests Erupt in City of Puno. BBC.com, May 27. http://www.bbc.co.uk/news/world-latin-america-13582707.

Bebbington, A. J., M. Connarty, W. Coxshall, H. O'Shaugnessy, and M. Williams. 2007. *Mining and Development in Peru, with Special Reference to the Rio Blanco Project, Piura.* London: Peru Support Group.

Beck, Ulrich. 2004. The Truth of Others: A Cosmopolitan Approach. *Common Knowledge* 10 (3): 430–449.

Beck, Ulrich. 2005. *Power in the Global Age: A New Political Economy.* Cambridge, UK: Polity Press.

Beck, Ulrich. 2010. A Cosmopolitan Manifesto. In *The Cosmopolitanism Reader*, ed. Garrett Wallace Brown and David Held. Cambridge, UK: Polity Press.

Bellman, Christophe, Graham Dutfield, and Ricardo Melendez-Ortiz. 2004. *Trading in Knowledge: Development Perspectives on TRIPS, Trade, & Sustainability.* London: Earthscan Publications.

Benkler, Yochai. 2006. *The Wealth of Networks: How Social Production Transforms Markets and Freedom.* New Haven, CT: Yale University Press.

Berlin, Brent, and Ann Eloise Berlin. 2003. NGOs and the Process of Prior Informed Consent in Bioprospecting Research: The Maya ICBG Project in Chiapas, Mexico. *International Social Science Journal* 55 (178): 629–638.

Bessen, James. 2002. What Good Is Free Software? In *Government Policy toward Open Source Software*, ed. R. Hahn. Washington, DC: Brookings Institution Press.

Biagioli, Mario. 1993. *Galileo, Courtier: The Practice of Science in the Culture of Absolutism.* Chicago: University of Chicago Press.

Biagioli, Mario, and Peter Gallison, eds. 2003. *Scientific Authorship: Credit and Intellectual Property in Science.* New York: Routledge.

Biehl, Joao. 2005. *Vita: Life in a Zone of Social Abandonment.* Berkeley: University of California Press.

Blom Hansen, Thomas, and Finn Stepputat. 2001. *States of Imagination: Ethnographic Explorations of the Postcolonial State.* Durham, NC: Duke University Press.

Blom Hansen, Thomas, and Finn Stepputat, eds. 2005. *Sovereign Bodies: Citizens, Migrants, and States in the Postcolonial World.* Princeton, NJ: Princeton University Press.

Bowker, Geoffrey. 2005. *Memory Practices in the Sciences.* Cambridge, MA: MIT Press.

Boyle, James. 1996. *Shamans, Software & Spleens: Law & the Construction of Information Society.* Cambridge, MA: Harvard University Press.

Boyle, James. 2010. *The Public Domain: Enclosing the Commons of the Mind.* New Haven, CT: Yale University Press.

Bray, Hiawatha. 2007. One Laptop per Child Orders Surge: Peru Wants 260,000 Machines, Mexican Billionaire Signs Up. *The Boston Globe,* December 1. http://www.boston.com/business/technology/articles/2007/12/01/one_laptop_per_child_orders_surge.

Brenner, Neil, and Roger Keil, eds. 2006. *The Global Cities Reader.* New York: Routledge.

Brown, Garrett Wallace, and David Held, eds. 2010. *The Cosmopolitanism Reader.* Cambridge, UK: Polity Press.

Brown, Michael. 2003. *Who Owns Native Culture?* Cambridge, MA: Harvard University Press.

Brush, Stephen. 1996. Whose Knowledge, Whose Genes, Whose Rights? In *Valuing Local Knowledge,* ed. S. Brush and D. Stabinsky. Washington, DC: Island Press.

Brush, Stephen. 1999. Bioprospecting the Public Domain. *Cultural Anthropology* 14 (4): 535–555.

Buonomano, Dean. 2011. *Brain Bugs: How the Brain's Flaws Shape Our Lives.* New York: W. W. Norton.

Burnett, Judith, Peter Senker, and Kathy Walker, eds. 2009. *The Myths of Technology: Innovation & Inequality.* New York: Peter Lang.

Burns, Axel, and Ben Eltham. 2009. Twitter-Free Iran: An Evaluation of Twitter's Role in Public Diplomacy and Information Operations in Iran's 2009 Election Crisis. In *Record of the Communications Policy & Research Forum 2009,* ed. F. Papandrea and M. Armstrong. Sydney, Australia: Network Insight Institute.

Callon, Michel. 1986. Some Elements of a Sociology of Translation: Domestication of the Scallops and the Fisherman of St. Brieuc Bay. In *Power, Action, and Belief: A New Sociology of Knowledge?* ed. J. Law. London: Routledge.

Callon, Michel, and Bruno Latour. 1981. Unscrewing the Big Leviathan: How Actors Macro-Structure Reality and How Sociologists Help Them to Do So. In *Advances in Social Theory and Methodology: Toward an Integration of Micro- and Macro Sociologies*, ed. K. Knorr-Cetina and A. Cicourel. London: Routledge & Kegan Paul.

Canclini, Nestor Garcia. 2001. *Consumers and Citizens: Globalization and Multicultural Conflicts*. Minneapolis: University of Minnesota Press.

Cardenas, Jose Arturo. 2011. Peru Anti-Mining Protests Halted until after Election. *AFP News*, May 31. http://www.google.com/hostednews/afp/article/ALeqM5hkRl7sR qEgkdfj3vfhfSesh8wj2A?docId=CNG.4731b3aba4df7bc97fe661f788f35488.1091.

Carr, Nicolas. 2010. *The Shallows: What the Internet Is Doing to Our Brains*. New York: W. W. Norton.

Carroll, Rory. 2007. Peasant Leader Killed as Protests Paralyse Peru. *The Guardian*, July 17. http://www.guardian.co.uk/world/2007/jul/17/rorycarroll.

Castells, Manuel. 1996. *The Information Age: Economy, Society, and Culture*. Oxford: Blackwell.

Castells, Manuel. 2003. *The Internet Galaxy: Reflections on the Internet, Business, and Society*. Oxford: Oxford University Press.

Castells, Manuel. 2009. *Communication Power*. Oxford: Oxford University Press.

Castells, Manuel. 2012. *Networks of Outrage and Hope: Social Movements in the Internet Age*. Malden, MA: Polity Press.

Castells, Manuel, Mireia Fernandez-Ardevol, Jack Linchuan Qiu, and Araba Sey. 2009. *Mobile Communication and Society: A Global Perspective*. Cambridge, MA: MIT Press.

Castillo Butters, Luis Jaime, Leena Bernuy Quiroga, and Pamela Lastes Dammert. 2005. Internationalization of Higher Education in Peru. In *Higher Education in Latin America: The International Dimension*, ed. H. De Wit, I. Jaramillo, J. Gacel-Avila, and J. Knight. Washington, DC: The World Bank.

Castree, Noel. 2003. Bioprospecting: From Theory to Practice (and Back Again). *Transactions of the Institute of British Geographers* 28 (1): 35–55.

Chakrabarty, Dipesh. 2000. *Provincializing Europe: Postcolonial Thought and Historical Difference*. Princeton, NJ: Princeton University Press.

Chan, Anita. 2004a. Coding Free Software, Coding Free States: Free Software Legislation and the Politics of Code in Peru. *Anthropological Quarterly* 77 (3): 531–545.

Chan, Anita. 2004b. Seeing Free Software, Seeing Free States: Free Software Legislation and the Politics of Code in Peru. *Anthropological Quarterly* 77: 3.

Charbonnier, Nicolas. 2009. OLPC's Netbook Impact on Laptop PC Industry. Olpc-news.com, December 7. www.olpcnews.com/commentary/impact/olpc_netbook_impact_on_laptop.html.

Cheah, Pheng. 1998. The Cosmopolitical, Today. In *Cosmopolitics: Thinking and Feeling beyond the Nation*, ed. P. Cheah and B. Robbins. Minneapolis: University of Minnesota Press.

Cheah, Pheng, and Bruce Robbins, eds. 1998. *Cosmopolitics: Thinking and Feeling beyond the Nation*. Minneapolis: University of Minnesota Press.

Christensen, Clayton. 2011. *The Innovator's Dilemma: The Revolutionary Book That Will Change the Way You Do Business*. New York: Harpers Press.

Clinton, Hillary. 2010. Remarks on Internet Freedom. Delivered at the Newseum, Washington, DC, January 21, 2010. http://www.state.gov/secretary/rm/2010/01/135519.htm.

Coleman, Gabriella. 2003. The Political Agnosticism of Free and Open Source Software and the Inadvertent Politics of Contrast. Paper presented at the American Anthropological Association, Chicago, November 22.

Coleman, Gabriella. 2004. The Political Agnosticism of Free and Open Source Software and the Inadvertent Politics of Contrast. *Anthropological Quarterly* 77 (3): 507–519.

Coleman, Gabriella. 2010. The Hacker Conference: A Ritual Condensation and Celebration of Lifeworld. *Anthropological Quarterly* 83 (1): 47–72.

Coleman, Gabriella. 2011. Anonymous: From the Lulz to Collective Action. *The New Everyday*, March. http://mediacommons.futureofthebook.org/tne/pieces/anonymous-lulz-collective-action.

Coleman, Gabriella. 2013. *Coding Freedom: Hacker Pleasure and the Ethics of Free and Open Source Software*. Princeton, NJ: Princeton University Press.

Coleman, Gabriella, and Mako Hill. 2004. How Free Became Open and Everything Else under the Sun. *media/culture* 7: 3.

Collyns, Dan. 2007. Street Protests "Paralyse" Peru. BBC.com, July 15. http://news.bbc.co.uk/2/hi/americas/6899331.stm.

Comaroff, Jean, and John Comaroff. 2001. *Millennial Capitalism and the Culture of Neoliberalism*. Durham, NC: Duke University Press.

Comaroff, Jean, and John Comaroff. 2005. Aliens, Apocalypse and the Postcolonial State. In *Sovereign Bodies: Citizens, Migrants, and States in the Postcolonial World*, ed. Thomas Blom Hansen and Finn Stepputat. Princeton, NJ: Princeton University Press.

Comaroff, Jean, and John Comaroff. 2009. *Ethnicity, Inc.* Chicago: University of Chicago Press.

Comision Multisectorial para el Desarrollo de la Sociedad de la Informacion (CODESI). 2005. *Plan de Desarrollo de la Sociedad de Informacion en el Peru: La Agenda Digital Peruana*. Lima: Oficina Nacional de Gobierno Electrónico e Informática.

Comision Multisectorial para el Desarrollo de la Sociedad de la Informacion (CODESI). 2011. *Plan de Desarrollo de la Sociedad de Informacion en el Peru: La Agenda Digital 2.0*. Lima: Oficina Nacional de Gobierno Electrónico e Informática.

Coombe, Rosemary. 1996. Embodied Trademarks: Mimesis and Alterity on American Commercial Frontiers. *Cultural Anthropology* 11 (2): 202–224.

Coombe, Rosemary. 1998a. *The Cultural Life of Intellectual Properties: Authorship, Appropriation and the Law*. Durham, NC: Duke University Press.

Coombe, Rosemary. 1998b. Intellectual Property, Human Rights and Sovereignty: New Dilemmas in International Law Posed by the Recognition of Indigenous Knowledge and the Convention on Biodiversity. *Indiana Journal of Global Legal Studies* 6 (1): 59–115.

Coombe, Rosemary. 2005. Protecting Traditional Environmental Knowledge and New Social Movements in the Americas: Intellectual Property, Human Right or Claims to an Alternative Form of Sustainable Development? *Florida Journal of International Law* 17 (1).

Coombe, Rosemary, Steven Schnoor, and Al Attar Ahmed Mohsen. 2000. *Intellectual Property Rights, the WTO and Developing Countries: The TRIPS Agreement and Policy Options*. New York: Zed Books.

Coombe, Rosemary, Steven Schnoor, and Al Attar Ahmed Mohsen. 2006. Bearing Cultural Distinction: Informational Capitalism and New Expectations for Intellectual Property. In *Articles in Intellectual Property: Crossing Borders*, ed. F. Grosheide and J. Brinkhof. Utrecht, The Netherlands: Mollengrafica/Intersentia.

Correa, Carlos. 2000. *Intellectual Property Rights, the WTO and Developing Countries: The TRIPS Agreement and Policy Options*. New York: Zed Books.

Correa, Carlos. 2003. The Access Regime and the Implementation of the FAO International Treaty on Plant Genetic Resources for Food and Agriculture in the Andean Group Countries. *Journal of World Intellectual Property* 6 (6): 795–806.

Cringley, Robert. 1996. *Accidental Empires: How the Boys of Silicon Valley Make Their Millions, Battle Foreign Competition, and Still Can't Get a Date*. New York: Harpers Press.

Darnton, Robert. 1999. *The Great Cat Massacre: And Other Episodes in French Cultural History*. New York: Vintage Books.

Dayan, Daniel, and Elihu Katz. 1994. *Media Events: The Live Broadcasting of History*. Cambridge, MA: Harvard University Press.

de Castro, Eduardo Batalha Viveiros. 2004. Exchanging Perspectives: The Transformation of Objects into Subjects in Amerindian Ontologies. *Common Knowledge* 10 (3): 463–484.

de la Cadena, Marisol. 2000. *Indigenous Mestizos: The Politics of Race and Culture in Cuzco, Peru, 1919–1991*. Durham, NC: Duke University Press.

de la Cadena, Marisol. 2010. Indigenous Cosmopolitics in the Andes: Conceptual Reflections beyond "Politics." *Cultural Anthropology* 25 (2): 334–370.

de la Pena, Carolyn Thomas. 2006. "Slow and Low Progress," or Why American Studies Should Do Technology. *American Quarterly* 58 (3): 915–941.

de Wit, Hans, Isabel Christina Jaramillo, Jocelyne Gacel-Avila, and Jane Knight, eds. 2005. *Higher Education in Latin America: The International Dimension*. Washington, DC: The World Bank.

Defensoria del Pueblo de la Republica del Peru. 2006. Informe Anual de la Defensoria del Pueblo 2006.

Defensoria del Pueblo de la Republica del Peru. 2007. Informe Anual de la Defensoria del Pueblo 2007.

Defensoria del Pueblo de la Republica del Peru. 2008. Informe Anual de la Defensoria del Pueblo 2008.

DeHart, Monica. 2010. *Ethnic Entrepreneurs: Identity and Development Politics in Latin America*. Palo Alto, CA: Stanford University Press.

Derndorfer, Cristoph. 2010. OLPC in Peru: A Problematic Una Laptop por Nino Program. EdutechDebate, October 28. http://edutechdebate.org/olpc-in-south-america/olpc-in-peru-one-laptop-per-child-problems.

Dibbell, Julian. 2007. *Play Money: Or How I Quit My Job and Made Millions Trading Virtual Loot*. New York: Basic Books.

Dorn, Matt. 2003. The Free Software Tango. Salon.com, May 7. http://www.salon.com/tech/feature/2003/05/07/free_software_argentina.

Douglas, Mary. 1966. *Purity and Danger: An Analysis of Concepts of Pollution and Taboo*. New York: Praeger.

Dourish, Paul, and Genevieve Bell. 2011. *Divining a Digital Future: Mess and Mythology in Ubiquitous Computing*. Cambridge, MA: MIT Press.

Dove, Michael. 1996. Center, Periphery, and Biodiversity: A Paradox of Governance and a Developmental Challenge. In *Valuing Local Knowledge*, ed. S. Brush and D. Stabinsky. Washington, DC: Island Press.

Drahos, Peter, and Ruth Mayne, eds. 2002. *Global Intellectual Property Rights: Knowledge, Access and Development*. New York: Palgrave McMillan.

Durant, Alberto. 2009. *Donde Esta El Pirata?: Para Entender el Comercio Informal de Peliculas Digitales en el Peru*. Lima, Peru: Remanso Ediciones EIRL.

Dussel, Enrique. 2000. Europe, Modernity and Eurocentrism. *Nepantla: Views from South* 1 (3): 465–478.

Dussel, Enrique. 2002. World-System and "Trans" Modernity. *Nepantla: Views from South* 2 (3): 221–245.

Dutfield, Graham. 2000. *Intellectual Property Rights, Trade, and Biodiversity*. London: Earthscan Publications.

Dutfield, Graham. 2002a. *Protecting Traditional Knowledge and Folklore: A Review of Progress in Diplomacy and Policy Formation*. Geneva: International Centre for Trade and Sustainable Development (ICTSD) International Environment House.

Dutfield, Graham. 2002b. Sharing the Benefits of Biodiversity: Is There a Role for the Patent System? *Journal of World Intellectual Property* 5 (6): 899–931.

Dutfield, Graham, and Uma Suthersanen. 2008. *Global Intellectual Property Law*. Cheltenham, UK: Edward Elgar.

Eckstein, Susan, and Manuel Merino, eds. 2001. *Power and Popular Protest: Latin American Social Movements*. Berkeley: University of California Press.

Economic Times. 2010. Incredible India: Game-Changing Low-Cost Laptop, July 24. http://articles.economictimes.indiatimes.com/2010-07-24/news/27604238_1_internet-access-pc-hard-disk.

Economist. 2008. One Clunky Laptop per Child, January 4.

El Comercio. 2009. Junio fue el mes de mayor numero de conflictos sociales en el ultimo ano. elcomercio.com.pe, July 8. http://www.elcomercio.com.pe/noticia/311188/junio-fue-mes-mayor-numero-conflictos-sociales-ultimo-ano.

El Comercio. 2011. Una Laptop Por Nino: Por una educacion de calidad con equidad. January 17.

Escobar, Arturo. 1994. *Encountering Development: The Making and Unmaking of the Third World*. Princeton, NJ: Princeton University Press.

Escobar, Arturo. 2007. Worlds and Knowledges Otherwise: The Latin American Modernity/Coloniality Research Program. *Cultural Studies* 21 (2): 179–210.

Escobar, Arturo. 2008. *Territories of Difference: Place, Movements, Life, Redes*. Durham, NC: Duke University Press.

Evans, David. 2002. Politics and Programming: Government Preferences for Open Source Software. In *Government Policy toward Open Source Software*, ed. R. Hahn. Washington, DC: Brookings Institution Press.

Faye, David. 2004. Bioprospecting, Genetic Patenting and Indigenous Populations: Challenges under a Restructured Information Commons. *Journal of World Intellectual Property* 7 (3): 401–428.

Feller, Joseph, Brian Fitzgerald, Scott Hissam, and Karim Lakhani. 2005. *Perspectives on Free and Open Source Software.* Cambridge, MA: MIT Press.

Ferguson, James. 1994. *Anti-Politics Machine: Development, Depoliticization, and Bureaucratic Power in Lesotho.* Minneapolis: University of Minnesota Press.

Ferrer, Edgar. 2009. ICT Policy and Perspectives of Human Development in Latin America: The Peruvian Experience. *Journal of Technology Management & Innovation* 4 (4): 162–170.

Ferris, Timothy. 2010. *The Science of Liberty: Democracy, Reason, and the Laws of Nature.* New York: Harper.

Festa, Paul. 2001. Governments Push Open Source Software. Cnet.com, August 29. http://news.com.com/2100-1001-272299.html?legacy=cnet.

Finger, J. Michael, and Peter Schuler. 2004. *Poor People's Knowledge: Promoting Intellectual Property in Developing Countries.* Washington, DC: World Bank Publications.

Florida, Richard. 2003. *The Rise of the Creative Class: And How It's Transforming Work, Leisure, Community, and Everyday Life.* New York: Basic Books.

Foucault, Michel. 1977. *Language, Counter-Memory, Practice: Selected Essays and Interviews by Michel Foucault.* Ithaca, NY: Cornell University Press.

Foucault, Michel. 2007. *Security, Territory, Population: Lectures at the College de France, 1977–78.* New York: Palgrave McMillan.

Fouche, Rayvon. 2012. From Black Inventors to One Laptop per Child: Exporting a Racial Politics of Technology. In *Race After the Internet,* ed. L. Nakamura and P. Chow. New York: Routledge.

Fox, Richard, and Orin Starn. 1997. *Between Resistance and Revolution: Cultural Politics and Social Protest.* New Brunswick, NJ: Rutgers University Press.

Friedman, Thomas. 2005. *The World Is Flat: A Brief History of the Twenty-First Century.* New York: Farrar, Straus and Giroux.

Gaines, Liz. 2009. TED: Negroponte Says OLPC Started Netbook Craze; Will Open-Source Its Hardware. Gigaom.com, February 7. http://gigaom.com/2009/02/07/ted-negroponte-says-olpc-started-netbook-craze-will-open-source-its-hardware.

Gervais, Daniel. 2003. TRIPS, Doha and Traditional Knowledge. *Journal of World Intellectual Property* 6 (3): 403–419.

Ghosh, R. 2005. Understanding Free Software Developers: Findings from the FLOSS Study. In *Perspectives on Free and Open Source Software,* ed. J. Feller, B. Fitzgerald, S. Hissam, and K. Lakhani. Cambridge, MA: MIT Press.

Ghosh, R., R. Glott, B. Krieger, and G. Robles. 2002. *Free/Libre and Open Source Software: Survey and Study.* Maastricht, The Netherlands: University of Maastricht, International Institute of Infonomics Publications.

Glieck, James. 2011. *The Information: A History, a Theory, a Flood.* New York: Pantheon.

Gluckman, Max. 1963. Gossip and Scandal. *Current Anthropology* 4: 307–316.

Gobierno de Peru. 2006. PENX 2003–2013: Master Plan for Export Culture. Lima, Peru.

Goodale, Mark, and Sally Engle Merry, eds. 2007. *The Practice of Human Rights: Tracking Law between the Global and Local.* Cambridge, UK: Cambridge University Press.

Graham, Laura. 2002. How Should an Indian Speak? Amazonian Indians and the Symbolic Politics of Language in the Global Public Sphere. In *Indigenous Movements, Self-Representation, and the State in Latin America,* ed. Kay Warren and Jean Jackson. Austin: University of Texas Press.

Greene, Thomas. 2002. MS in Peruvian Open-Source Nightmare. TheRegister.com, May 5. http://theregister.co.uk/content/archive/25157.html.

Grossman, Lev. 2009. Iran's Protests: Why Twitter Is the Medium of the Movement. *Time Magazine,* June 17.

Guerrero, Carlos. 2011. Mine Opponents Paralyze City in Peruvian Highlands. Associated Press, May 27. http://www.msnbc.msn.com/id/43193125/ns/world_news-americas/t/mine-opponents-paralyze-city-peruvian-highlands.

Habermas, Jürgen. 2010. A Political Constitution for the Pluralist World Society? In *The Cosmopolitanism Reader,* ed. Garrett Wallace Brown and David Held. Cambridge, UK: Polity Press.

Hackworth, Jason. 2006. *The Neoliberal City: Governance, Ideology, and Development in American Urbanism.* Ithaca, NY: Cornell University Press.

Hafner, Katie, and Matthew Lyon. 1996. *Where Wizards Stay up Late: The Origins of the Internet.* New York: Simon & Schuster.

Hahn, Robert, ed. 2002. *Government Policy toward Open Source Software.* Washington, DC: Brookings Institution Press.

Halliday, Josh. 2011. Hillary Clinton Adviser Compares Internet to Che Guevara. guardian.com, June 22, 2011. http://www.guardian.co.uk/media/2011/jun/22/hillary-clinton-adviser-alec-ross.

Haraway, Donna. 1991. Situated Knowledges: The Science Question in Feminism and the Privilege of Partial Perspective. In *Simians, Cyborgs and Women: The Reinvention of Nature.* New York: Routledge.

Haraway, Donna. 1997. *Modest_Witness@Second_Millenium: FemaleMan_Meets_Oncomouse*. New York: Routledge.

Harding, Sandra. 1998. *Is Science Multicultural?: Postcolonialisms, Feminisms, and Epistemologies*. Indianapolis: Indiana University Press.

Harding, Sandra. 2008. *Science from Below: Feminisms, Postcolonialities, and Modernities*. Chapel Hill, NC: Duke University Press.

Harris, Olivia. 1995. The Sources and Meanings of Money: Beyond the Market Paradigm in an Ayllu of Northern Potosi. In *Ethnicity, Markets, and Migration in the Andes: At the Crossroads of History and Anthropology*, ed. Brook Larson and Olivia Harris. Durham, NC: Duke University Press.

Harvey, David. 2005. *A Brief History of Neoliberalism*. Oxford: Oxford University Press.

Harvey, David. 2006. *Spaces of Global Capitalism: A Theory of Uneven Geographical Development*. New York: Verso.

Hayden, Cori. 2003. *When Nature Goes Public: The Making and Unmaking of Bioprospecting in Mexico*. Princeton, NJ: Princeton University Press.

Hilgartner, S. 2009. Intellectual Property and the Politics of Emerging Technology. *Chicago-Kent Law Review* 84 (1): 197–224.

Hiltzik, Michael. 2000. *Dealers of Lightning: Xerox PARC and the Dawn of the Computer Age*. New York: Harper Books.

Himanen, Pekka. 2001. *The Hacker Ethic and the Spirit of Capitalism*. New York: Random House.

Howkins, John. 2002. *The Creative Economy: How People Make Money from Ideas*. London: Penguin Press.

INDECOPI. 2010. *Denominacion de Origen: Guia Informativa*. Lima, Peru: Centro de Informacion y Documentacion de INDECOPI.

Instituto Nacional de Estatistica e Informatica (INEI). 2002. *Plan de Implementacion de Software Libre en la Administracion Publica*.

International Labor Organization [ILO]. 1989. Indigenous and Tribal Peoples Convention.

International Monetary Fund. 2011. Public Notice No. 11/158. http://www.imf.org/external/np/sec/pn/2011/pn11158.htm.

Isaac, Grant, and William Kerr. 2004. Bioprospecting or Biopiracy?: Intellectual Property and Traditional Knowledge in Biotechnology Innovation. *Journal of World Intellectual Property* 7 (1): 35–52.

Jackson, Jean. 2007. Rights to Indigenous Culture in Colombia. In *The Practice of Human Rights: Tracking Law between the Global and Local*, ed. Mark Goodale and Sally Engle Merry. Cambridge, UK: Cambridge University Press.

James, Jeffrey. 2010. New Technology in Developing Countries: A Critique of the One-Laptop-Per-Child Program. *Social Science Computer Review* 28 (3): 381–390.

Jarvis, Jeff. 2009. *What Would Google Do?* New York: Harpers Press.

Jenkins, Henry. 2008. *Convergence Culture: Where Old and New Media Collide*. New York: New York University Press.

Johnson, Bobbie. 2011. How Twitter Could Bring About World Peace. Gigaom .com, April 11. http://gigaom.com/2011/04/11/how-twitter-could-bring-about -world-peace.

Johnson, Steven. 2010. *Where Good Ideas From: The Natural History of Innovation*. New York: Riverhead.

Joseph, Gilbert, and David Nugent, eds. 1994a. *Everyday Forms of State Formation: Revolution and the Negotiation of Rule in Modern Day Mexico*. Durham, NC: Duke University Press.

Joseph, Gilbert, and David Nugent. 1994b. Popular Culture and State Formation in Revolutionary Mexico. In *Everyday Forms of State Formation: Revolution and Negotiation of Rule in Modern Mexico*, ed. Gilbert Joseph and David Nugent. Durham, NC: Duke University Press.

Kant, Immanuel. 1784. The Idea of a Universal History with a Cosmopolitan Intent. In *Perpetual Peace and Other Essays on Politics, History, and Morals*, ed. Ted Humphrey. Indianapolis: Hackett Publishing.

Katz, Leslie. 2010. India's $35 Tablet—How Low Can It Go? CNET.com, July 23. http://news.cnet.com/8301-17938_105-20011536-1.html.

Keck, Margaret, and Kathryn Sikkink. 1998. *Activists beyond Borders: Advocacy Networks in International Politics*. Ithaca, NY: Cornell University Press.

Kelly, Kevin. 2010. *What Technology Wants*. New York: Viking Books.

Kelty, Chris. 2001. Free Software/Free Science. *First Monday* 6: 12.

Kelty, Chris. 2005a. Free Science. In *Perspectives on Free and Open Source Software*, ed. J. Feller, B. Fitzgerald, S. Hissam, and K. Lakhani. Cambridge, MA: MIT Press.

Kelty, Chris. 2005b. Geeks, Social Imaginaries, and Recursive Publics. *Cultural Anthropology* 20 (2): 185–214.

Kelty, Chris. 2008. *Two Bits: The Cultural Significance of Free Software*. Chapel Hill, NC: Duke University Press.

Kidder, Tracy. 2000. *The Soul of a New Machine*. Boston: Little, Brown.

Klein, Naomi. 2007. *Shock Doctrine: The Rise of Disaster Capitalism*. New York: Metropolitan Books.

Klein, S. J. 2011. TEDx Rio: Rodrigo on OLPC and Viewing Children as Our Future. *The OLPC Blog*. http://blog.laptop.org/2011/02/16/tedxrio-rodrigo-olpc.

Kraemer, Kenneth, Jason Dedrick, and Pauline Sharma. 2009. One Laptop per Child: Vision vs. Reality. *Communications of the ACM* 52: 6.

Krikorian, Gaëlle, and Amy Kapczynski, eds. 2010. *Access to Knowledge in the Age of Intellectual Property*. New York: Zone Books.

Lacan, Jacques. 1998. The Subject and the Other: Alienation. In *The Four Fundamental Concepts of Psychoanalysis*, ed. Jacques-Alain Miller. New York: W. W. Norton.

Lacan, Jacques. 2002. The Mirror Stage as Formative of the *I* Function as Revealed in Psychoanalytic Experience. In *Ecrits: A Selection*. New York: W. W. Norton.

LACFREE (Latin American and Caribbean Conference on Free Software Development and Use). 2003. Carta de Cusco. http://www.somoslibres.org/modules.php?name=Content&pa=showpage&pid=10.

Lakhani, Karim, and Robert Wolf. 2005. Why Hackers Do What They Do: Understanding Motivation and Effort in Free/Open Source Software Projects. In *Perspectives on Free and Open Source Software*, ed. J. Feller, B. Fitzgerald, S. Hissam, and K. Lakhani. Cambridge, MA: MIT Press.

Lanier, Jaron. 2010. *You Are Not a Gadget: A Manifesto*. New York: Knopf.

Latour, Bruno. 1987. *Science in Action: How to Follow Scientists and Engineers through Society*. Cambridge, MA: Harvard University Press.

Latour, Bruno. 1990. Drawing Things Together. In *Representation in Scientific Practice*, ed. M. Lynch and S. Woolgar, 19–68. Cambridge, MA: MIT Press.

Latour, Bruno. 1993. *We Have Never Been Modern*. Cambridge, MA: Harvard University Press.

Latour, Bruno. 2004a. *Politics of Nature: How to Bring the Sciences into Democracy*. Cambridge, MA: Harvard University Press.

Latour, Bruno. 2004b. Whose Cosmos, Which Cosmopolitics? Comments on the Peace Terms of Ulrich Beck. *Common Knowledge* 10 (3): 450–462.

Latour, Bruno. 2005. *Reassembling the Social: An Introduction to Actor-Network-Theory*. Oxford: Oxford University Press.

Latour, Bruno. 2010. *On the Modern Cult of the Factish Gods*. Durham, NC: Duke University Press.

Law, John. 2007. Pinboards and Books: Juxtaposing, Learning, and Materiality. In *Education and Technology: Critical Perspectives, Possible Futures*, ed. David Kritt and Lucien Winegar. New York: Lexington Books.

LeBlanc, Denise. 2002. Ending Microsoft FUD: An Interview with Peruvian Congressman Villanueva. LinuxToday.com, May 21. http://linuxtoday.com/news_story .php3?ltsn=2002–05–20–006–26-IN-LF-PB.

Lees, Loretta, Tom Salter, and Elvin Wyly. 2007. *Gentrification*. New York: Routledge.

Lessig, Lawrence. 2000. *Code and Other Laws of Cyberspace*. New York: Basic Books.

Lessig, Lawrence. 2002. *The Future of Ideas: The Fate of the Commons in a Connected World*. New York: Vintage.

Lessig, Lawrence. 2005. *Free Culture: The Nature and Future of Creativity*. New York: Penguin Press.

Lessig, Lawrence. 2008. *Remix: Making Art and Commerce Thrive in the Hybrid Economy*. New York: Penguin Press.

Lettice, John. 2002. Peru Mulls Free Software, Gates Gives $550k to Peru Prez. *TheRegister.com*, July 16. http://www.theregister.co.uk/content/4/26207.html.

Levy, Steven. 1984. *Hackers: Heroes of the Computer Revolution*. New York: Penguin.

Levy, Steven. 2000. *Insanely Great: The Life and Times of Macintosh, the Computer That Changed Everything*. New York: Penguin Press.

Levy, Steven. 2011. *In the Plex: How Google Thinks, Works, and Shapes Our Lives*. New York: Simon & Schuster.

Lewen, U. 2008. Internet File-Sharing: Swedish Pirates Challenge the U.S. *Cardozo Journal of International and Comparative Law* 16: 173–206.

Lewis, Michael. 2001. *The New New Thing*. New York: Penguin Books.

Li, Miaoran. 2009. The Pirate Party and the Pirate Bay: How the Pirate Bay Influences Swede and International Copyright Relations. *Pace International Law Review* 21: 281–307.

Liang, Lawrence. 2003a. Piracy, Creativity and Infrastructure: Rethinking Access to Culture. http://www.altlawforum.org/intellectual-property/publications/articles-on -the-social-life-of-media-piracy/Liang-Piracy-%20Infrastructure%20and%20Creativity .pdf/view.

Liang, Lawrence. 2003b. *Pirate Aesthetics*. Magazine of the Steirischer Herbst 2006 Festival at Graz. http://www.altlawforum.org/intellectual-property/publications/ articles-on-the-social-life-of-media-piracy/pirate-aesthetics.

Liang, Lawrence. 2004. Media Empires and Renegade Pirates. *Humanscape* 11: 8.

Liang, Lawrence. 2005. Porous Legalities and Avenues of Participation. *Sarai Reader* 2005: 6–17.

Liang, Lawrence, and Achal Prabhala. 2007. Reconsidering the Pirate Nation. Presented at Yale University Law School, Yale Information Society Project. Access to Knowledge Conference, April 27–29.

Lilputing.com. 2008. Comprehensive List of Low-Cost Ultraportables. http://liliputing.com/2008/04/over-past-six-months-or-so-asus-everex_24.html.

Lin, Yuwei. 2007. Hacker Culture and the FLOSS Innovation. In *Handbook on Research in Open Source Software: Technological, Economic, and Social Perspectives*, ed. K. St. Amant and B. Still. Hershey, PA: Idea Group.

Living in Peru. 2007. SUTEP Strike Takes Peru by Storm. LIP.com, July 5. http://www.livinginperu.com/news/4204.

MacKinnon, Rebecca. 2012. *Consent of the Networked: The Worldwide Struggle for Internet Freedom*. New York: Basic Books.

Mallon, Florencia. 1994. Reflection on the Ruins. In *Everyday Forms of State Formation: Revolution and Negotiation of Rule in Modern Mexico*, ed. Gilbert Joseph and Daniel Nugent. Durham, NC: Duke University Press.

Mallon, Florencia. 1995. *Peasant and Nation: The Making of Postcolonial Mexico and Peru*. Berkeley: University of California Press.

Mariátegui, José Carlos. 1988. *Seven Essays of the Peruvian Reality*. Austin: University of Texas Press.

Markoff, John. 2006. *What the Dormouse Said: How the Sixties Counterculture Shaped the Personal Computer Industry*. New York: Penguin Books.

Maskus, Keith. 2000. *Intellectual Property Rights in the Global Economy*. Washington, DC: Institute for International Economics.

Massey, Doreen. 2007. *World City*. New York: Polity Press.

Matthews, Duncan. 2002. *Globalizing Intellectual Property Rights: The TRIPS Agreement*. London: Routledge.

Mazzotti, Jose Antonio. 2008. Creole Agencies and the (Post)Colonial Debate in Spanish America. In *Coloniality at Large: Latin America and the Postcolonial Debate*, ed. Mabel Morana, Enrique Dussel, and Carlos Jauregui. Durham, NC: Duke University Press.

McGonigal, Jane. 2011. *Reality Is Broken: Why Games Make Us Better and How They Can Change the World*. New York: Penguin Books.

Medina, Eden, Ivan da Costa, and Christina Holmes. 2012. Beyond Imported Magic: Studying Science and Technology in Latin America. Paper presented for Beyond Imported Magic Workshop, University of Indiana, Bloomington, August 24.

Melucci, Alberto. 1996. *Challenging Codes: Collective Action in the Information Age.* Cambridge, UK: Cambridge University Press.

Merson, John. 2000. Bio-Prospecting or Bio-Piracy: Intellectual Property Rights and Biodiversity in a Colonial and Post-Colonial Context. *Osiris* 15: 282–296.

Microsoft. 2002. Peru Begins a New Year of Technological Modernization through Accord Signing between President Alejandro Toledo and Bill Gates. http://www.microsoft.com/peru/convenioperu.

Miegel, Fredrik, and Tobias Olsson. 2008. From Pirates to Politicians: The Story of the Swedish File Sharers Who Became a Political Party. In *Democracy, Journalism, and Technology: New Developments in an Enlarged Europe*, ed. N. Carpentier, P. Pruulmann-Vengerfeldt, K. Nordenstreng, M. Hartmann, P. Vihalemm, B. Cammaerts, H. Neiminen, and T. Olsson. Tartu, Estonia: Tartu University Press.

Mignolo, Walter. 2007. Delinking: The Rhetoric of Modernity, the Logic of Coloniality and the Grammar of De-coloniality. *Cultural Studies* 21 (2): 449–514.

Miller, Daniel, and Don Slater. 2001. *The Internet: An Ethnographic Approach.* London: Berg Publishers.

Mitchell, Timothy. 1990. *Colonizing Egypt.* Berkeley: University of California Press.

Mitchell, Timothy. 2002. *Rule of Experts: Egypt, Techno-Politics, Modernity.* Berkeley: University of California Press.

Mitchell, Timothy. 2005. The Work of Economics: How a Discipline Makes Its World. *European Journal of Sociology* 45 (2): 297–320.

Montejo, Victor. 2002. The Multiplicity of Mayan Voices: Mayan Leadership and the Politics of Self-Representation. In *Indigenous Movements, Self-Representation, and the State in Latin America*, ed. Kay Warren and Jean Jackson. Austin: University of Texas Press.

Moody, Glyn. 2002. *Rebel Code: Linux and the Open Source Revolution.* New York: Basic Books.

Mooney, Chris. 2005. *The Republican War on Science.* New York: Perseus.

Morana, Mabel, Enrique Dussel, and Carlos Jauregui, eds. 2008. *Coloniality at Large: Latin America and the Postcolonial Debate.* Durham, NC: Duke University Press.

Morozov, Evgeny. 2010a. *The Net Delusion: The Dark Side of Internet Freedom.* New York: Public Affairs.

Morozov, Evgeny. 2010b. Technological Utopianism. *Boston Review* 35: 6. http://bostonreview.net/BR35.6/ndf_technology.php.

Najambadi, Afsaneh. 1991. Interview with Gayatri Spivak. *Social Text* 9(3): 122–134.

Nash, June. 1993. *We Eat the Mines and the Mines Eat Us: Dependency and Exploitation in Bolivian Tin Mines*. New York: Columbia University Press.

Negroponte, Nicolas. 2009. Lessons Learned and Future Challenges. Speech, Reinventing the Classroom: Social and Educational Impact of Information and Communication Technologies in Education Forum, Washington, DC, September 15. http://www.olpctalks.com/nicholas_negroponte.

Negroponte, Nicolas. 2010a. Laptops Work. *Boston Review* 35 (6). http://bostonreview.net/BR35.6/ndf_technology.php.

Negroponte, Nicolas. 2010b. Open Letter to India, July 29. http://laptop.org/en/vision/essays/35-tablet.shtml.

Negroponte, Nicolas. 2012. "Teaching Children to Learn." Speech delivered at 2012 EmTech MIT: Emerging Technology MIT conference, Cambridge, MA. https://www2.technologyreview.com/emtech/12/video/#!/watch/nicholas-negroponte-teaching-children-to-learn.

Niezen, Ronald. 2003. *The Origins of Indigenism: Human Rights and the Politics of Identity*. Berkeley: University of California Press.

Nigh, Ronald. 2002. Maya Medicine in the Biological Gaze: Bioprospecting Research as Herbal Fetishism. *Current Anthropology* 43 (3): 451–477.

Nugent, David. 1997. *Modernity at the Edge of Empire: State, Individual, & the Nation in the Northern Peruvian Andes, 1885–1935*. Stanford, CA: Stanford University Press.

Nugent, David. 2001. Before History and Prior to Politics: Space, Time, and Territory in the Modern Peruvian Nation-State. In *States of Imagination: Ethnographic Explorations of the Postcolonial State*, ed. Thomas Blom Hansen and Finn Stepputat. Durham, NC: Duke University Press.

Nussbaum, Martha. 2010. Kant & Cosmopolitanism. In *The Cosmopolitanism Reader*, ed. Garrett Wallace Brown and David Held. Cambridge, UK: Polity Press.

Ong, Aiwah. 1999. *Flexible Citizenship: The Cultural Logics of Transnationalism*. Durham, NC: Duke University Press.

Ong, Aiwah. 2005. Splintering Cosmopolitanism: Asian Immigrants and Zones of Autonomy in the American West. In *Sovereign Bodies: Citizens, Migrants, and States in the Postcolonial World*, ed. Thomas Blom Hansen and Finn Stepputat. Princeton, NJ. Princeton University Press.

Ong, Aiwah. 2006. *Neoliberalism as Exception: Mutations in Citizenship and Sovereignty*. Durham, NC: Duke University Press.

Ong, Aiwah, and Stephen Collier, eds. 2004. *Global Assemblages: Technology, Politics, and Ethics as Anthropological Problems.* Malden, MA: Blackwell Publishing.

Oreskes, Naomi, and Eric Conway. 2010. *Merchants of Doubt: How a Handful of Scientists Obscured the Truth on Issues from Tobacco Smoke to Global Warming.* New York: Bloomsbury Press.

Parry, Bronwyn. 2000. The Fate of the Collections: Social Justice and the Annexation of Plant Genetic Resources. In *People, Plants and Justice: The Politics of Nature Conservation,* ed. C. Zerner. New York: Columbia University Press.

Parry, Bronwyn. 2002. Cultures of Knowledge: Investigating Intellectual Property Rights and Relations in the Pacific. *Antipode* 34 (4): 679–706.

Patry, William. 2009. *Moral Panics and the Copyright Wars.* Oxford: Oxford University Press.

Paul, Ryan. 2008. Classmate PC Gets a Boost with Million-Unit Venezuelan Order. Arstechnica.com, September 29.

Paul, Ryan. 2009. Behind the OLPC Layoffs: G1G1 Failure and Reduced Sponsorship. *Ars Technica,* January 30.

Paul, Ryan. 2010. OLPC's Negroponte Offers to Help India Realize $35 Tablet. Arstechnica.com, August 2010. help-india-realize-35-tablet.ars.

Peru Support Group. 2007. Teachers Protest over the Government's New Education Law. *Peru Support Group Publication,* 122.

Pfleifle, Mark. 2009. A Nobel Peace Prize for Twitter? *Christian Science Monitor,* July 6. http://www.csmonitor.com/Commentary/Opinion/2009/0706/p09s02-coop.html.

Philip, Kavita. 2005. What Is a Technological Author: The Pirate Function and Intellectual Property. *Postcolonial Studies* 8 (2): 199–218.

Philip, Kavita, Lily Irani, and Paul Dourish. 2010. Postcolonial Computing: A Tactical Survey. *Science, Technology, & Human Values.* http://www.humanities.uci.edu/critical/kp/pdf/Postcolonial_Computing_STHV2010.pdf.

Poole, Deborah. 1997. *Vision, Race, & Modernity: A Visual Economy of the Andean Image World.* Princeton, NJ: Princeton University Press.

Pooley, Eric. 2010. *The Climate War: True Believers, Power Brokers, and the Fight to Save the Earth.* New York: Hyperion.

Posey, Darrell, and Graham Dutfield. 1996. *Beyond Intellectual Property: Toward Traditional Resource Rights for Indigenous Peoples and Local Communities.* Ottawa, Canada: International Development Research Centre.

Povinelli, Elizabeth. 2002. *The Cunning of Recognition: Indigenous Alterities and the Making of Australian Multiculturalism.* Durham, NC: Duke University Press.

Prakash, Siddhartha. 1999. Towards a Synergy between Biodiversity and Intellectual Property Rights. *Journal of World Intellectual Property* 2 (5): 821–831.

Quijano, Anibal. 1993. América Latina en la economía mundial. *Problemas del desarollo* 24 (95).

Quijano, Anibal. 2000. Coloniality of Power, Eurocentrism, and Latin America. *Nepantla: Views from South* 1 (3): 533–580.

Quijano, Anibal. 2007. Coloniality and Modernity/Rationality (1989). reprinted in *Cultural Studies* 21 (2): 168–178.

Rama, Angel. 1996. *The Lettered City*. Durham, NC: Duke University Press.

Rheingold, Howard. 2003. *Smart Mobs: The Next Social Revolution*. New York: Basic Books.

Ribiero, John. 2010. India to Provide $35 Computing Device to Students. *Business Week*. July 22. http://www.businessweek.com/idg/2010-07-22/india-to-provide-35 -computing-device-to-students.html.

Riles, Annalise. 2001. *The Network Inside Out*. Ann Arbor: University of Michigan Press.

Riley, Mary. 2004. *Indigenous Intellectual Property Rights: Legal Obstacles and Innovative Solutions*. Lanham, MD: AltaMira Press.

Robbins, Bruce. 1998. Actually Existing Cosmopolitanism. In *Cosmopolitics: Thinking and Feeling beyond the Nation*, ed. P. Cheah and B. Robbins. Minneapolis: University of Minnesota Press.

Rockwell, Elsie. 1994. Schools of the Revolution: Enacting and Contesting State Forms in Tlaxcala. In *Everyday Forms of State Formation: Revolution and Negotiation of Rule in Modern Mexico*, ed. Gilbert Joseph and Daniel Nugent. Durham, NC: Duke University Press.

Rose, Nikolas. 1999. *The Powers of Freedom: Reframing Political Thought*. Cambridge, UK: Cambridge University Press.

Ross, Andrew. 2004. *No-Collar: The Humane Workplace and Its Hidden Costs*. Philadelphia: Temple University Press.

Ross, Andrew. 2010. *Nice Work If You Can Get It: Life and Labor in Precarious Times*. New York: New York University Press.

Said, Edward. 1979. *Orientalism*. New York: Vintage Books.

Salas, Sdenka. 2009. *La Laptop XO en el Aula*. Puno, Peru: Sagitario Impresores.

Salazar, Milagros. 2007. Escalating Conflicts Put Pres. García on the Spot. Inter Press News Service Agency, July 12. http://www.ipsnews.net/news.asp?idnews=38521.

Santiago, A., E. Severin, J. Cristia, P. Ibarrarán, J. Thompson, and S. Cueto. 2010. Evaluacíon Experimental del Programa "Una Laptop por Niño" en Perú. Washington, DC: Banco Interamericano de Desarrollo. http://www.iadb.org/document. cfm?id=35370099.

Sassen, Saskia. 1991. *The Global City: New York, London, Tokyo*. Princeton, NJ: Princeton University Press.

Sassen, Saskia. 2002. *Global Networks, Linked Cities*. New York: Routledge.

Sassen, Saskia. 2006. *Territory, Authority, Rights: From Medieval to Global Assemblages*. Princeton, NJ: Princeton University Press.

Saxenian, AnnaLee. 1996. *Regional Advantage: Culture and Competition in Silicon Valley and Route 128*. Cambridge, MA: Harvard University Press.

Scheeres, Julia. 2002. Peru Discovers Machu Penguin. Wired.com, April 22. http://wired.com/news/business/0,1367,51902,00.html.

Schulz, Markus. 2007. The Role of the Internet in Transnational Mobilization: A Case Study of the Zapatista Movement, 1994–2005. In *Civil Society: Local and Regional Responses to Global Challenges*, ed. Mark Herkenrath. Piscataway, NJ: Transaction Publishers.

Scott, Allan. 2001. *Global City-Regions: Trends, Theory, Policy*. Oxford: Oxford University Press.

Sell, Susan. 2003. *Private Power, Public Law: The Globalization of Intellectual Property Rights*. Cambridge, UK: Cambridge University Press.

Shapin, Steven. 1995. *A Social History of Truth: Civility and Science in Seventeenth-Century England*. Chicago: University of Chicago Press.

Shapin, Steven, and Simon Schaffer. 1985. *Leviathan and the Air Pump: Hobbes, Boyle, and the Experimental Life*. Princeton, NJ: Princeton University Press.

Shirky, Clay. 2009. *Here Comes Everybody: The Power of Organizing without Organizations*. New York: Penguin.

Shiva, Vandana. 1997. *Biopiracy: The Plunder of Nature and Knowledge*. Boston: South End Press.

Shiva, Vandana. 2004. Trips, Human Rights and the Public Domain. *Journal of World Intellectual Property* 7 (5): 665–673.

Simmel, Georg. 1990. *The Philosophy of Money*. London: Routledge.

Simmonds, Roger, and Gary Hack, eds. 2000. *Global City Regions: Their Emerging Forms*. New York: Routledge.

Slashdot.org. 2002a. Free Software Law in Peruvian Congress, May 4. http://slashdot.org/article.pl?sid=02/05/04/220237&mode=thread&tid=117.

Slashdot.org. 2002b. Peruvian Congressman vs. Microsoft FUD, May 6. http://slashdot.org/article.pl?sid=02/05/06/1739244&mode=thread&tid=109.

Smith, Barbara Herrnstein. 2011. The Chimera of Relativism: A Tragicomedy. *Common Knowledge* 17 (1): 13–28.

Smith, Merritt Roe, and Leo Marx. 1994. *Does Technology Drive History? The Dilemma of Technological Determinism*. Cambridge, MA: MIT Press.

Sniderman, Zachary. 2011. Intel's Little Laptop That Could Bring Tech to Millions of Children around the World. Mashable.com, July 5. http://mashable.com/2011/07/05/intel-classmate-pc/.

Soon-Hong, Choi. 2011. The Role of the UN: ICTs in Crisis Response, Peacekeeping, and Peacebuilding. In *Peacebuilding in the Information Age: Sifting Hype from Reality*, ed. ICT4Peace, The Berkman Center for Internet and Society, and The Georgia Institute of Technology. http://ict4peace.org/wp-content/uploads/2011/01/Peacebuilding-in-the-Information-Age-Sifting-Hype-from-Reality.pdf

Speed, Shannon. 2007. Exercising Rights and Reconfiguring Resistance in the Zapatista Juntas de Buen Gobierno. In *The Practice of Human Rights: Tracking Law between the Global and Local*, ed. Mark Goodale and Sally Engle Merry. Cambridge, UK: Cambridge University Press.

Stanco, Tony. 2003. Opinion on Brazil Making Open Source Mandatory in Government. *Linux Today*, July 13. http://linuxtoday.com/news_storyphp3?ltsn=2003-06-13-009-26-OS-LL-PB.

Star, Susan Leigh, and James Griesemer. 1999. Institutional Ecology, "Translations," and Boundary Objects. In *The Science Studies Reader*, ed. Mario Biagioli, 505–524. New York: Routledge.

Stark, David. 2001. Heterarchy: Exploiting Ambiguity and Organizing Diversity. *Brazilian Journal of Political Economy* 21 (1): 2–79.

Stark, David. 2009. *The Sense of Dissonance: Accounts for Worth in Economic Life*. Princeton, NJ: Princeton University Press.

Stauffacher, Daniel, William Drake, Paul Currion, and Julia Steinberger, eds. 2005. *Information and Communication Technology for Peace: The Role of ICT in Preventing, Responding to and Recovering from Conflict*. New York: United Nations Information and Communication Technologies Task Force.

Stengers, Isabelle. 2007. *Cosmopolitics*. St. Paul: University of Minnesota Press.

Stengers, Isabelle. 2011. Comparison as a Matter of Concern. *Common Knowledge* 17 (1): 48–63.

Stepputat, Finn. 2005. Violence, Sovereignty, and Citizenship in Postcolonial Peru. In *Sovereign Bodies: Citizens, Migrants, and States in the Postcolonial World*, ed. Thomas Blom Hansen and Finn Stepputat. Princeton, NJ: Princeton University Press.

Stiglitz, Joseph. 2010. *Freefall: America, Free Markets, and the Sinking of the World Economy*. New York: W. W. Norton.

Stocking, B. 2003 Vietnam Embracing Open-Source. *San Jose Mercury News*. http://www.siliconvalley.com/mld/siliconvalley/business/columnists/gmsv/7139304.htm.

Sugar, David. 2005. Free Software and Latin America: The Challenge Is Often Political Rather Than Technical. *Free Software Magazine*. http://www.freesoftwaremagazine.com/articles/latin_america.

Sundaram, Ravi. 1999. Recycling Modernity: Pirate Electronic Cultures in India. *Third Text* 13 (47).

Sunder Rajan, Kaushik. 2007. *Biocapital: The Constitution of Postgenomic Life*. Durham, NC: Duke University Press.

Suthersanen, Uma, Graham Dutfield, and Kit Boey Chow, eds. 2007. *Innovation without Patents: Harnessing the Creative Spirit of a Diverse World*. Northampton, MA: Ashgate Publishing.

Takhteyev, Yuri. 2012. *Coding Places: Software Practices in a South American City*. Cambridge, MA: MIT Press.

Talbot, David. 2008. Una Laptop por Nino. MIT Technology Review. April 22. http://www.technologyreview.com/featuredstory/409939/una-laptop-por-ni%C3%B1o.

Talbot, David. 2012. Given Tablets but No Teachers, Ethiopian Children Teach Themselves. MIT Technology Review. October 29. http://www.technologyreview.com/news/506466/given-tablets-but-no-teachers-ethiopian-children-teach-themselves.

Taussig, Michael. 1991. *Shamanism, Colonialism, and the Wild Man: A Study in Terror and Healing*. Chicago: University of Chicago Press.

Taussig, Michael. 2010. *The Devil and Commodity Fetishism in South America*. Chapel Hill: University of North Carolina Press.

Taylor, Peter. 2003. *World City Network: A Global Urban Analysis*. New York: Routledge.

Terranova, Tiziana. 2004. *Network Culture: Politics for the Information Age*. New York: Pluto Press.

Than, Ker. 2010. Internet Nominated for 2010 Nobel Peace Prize. MSNBC.com, March 11. http://www.msnbc.msn.com/id/35823790/ns/technology_and_science-tech_and_gadgets/t/internet-nominated-nobel-peace-prize.

Times of India. 2011. Where Is India's $35 Tablet, Mr Sibal? July 21. http://articles .timesofindia.indiatimes.com/2011-07-21/hardware/29799077_1_tablet-classmate -pc-laptop.

Toyama, Kentaro. 2010. Can Technology End Poverty? *Boston Review* 35 (6): 12–18, 28–29. http://bostonreview.net/BR35.6/ndf_technology.php.

Toyama, Kentaro. 2011. There Are No Technology Shortcuts to Good Education. EdutechDebate.org. http://edutechdebate.org/ict-in-schools/there-are-no-technology -shortcuts-to-good-education.

Tsing, Anna. 2004. *Friction: An Ethnography of Global Connection.* Princeton, NJ: Princeton University Press.

Tsotsis, Alexia. 2011. Internet up for Nobel Peace Prize Again, Let's Hope It Wins This Time. techcrunch.com, March 1. http://techcrunch.com/2011/03/01/internet -for-the-nobel-win.

Turkle, Sherry. 1995. *Life on the Screen: Identity in the Age of the Internet.* New York: Touchstone Books.

Turkle, Sherry. 2004. Wither Psychoanalysis in Digital Culture? *Psychoanalytic Psychology* 21: 16–30.

Turkle, Sherry. 2009. *Simulation and Its Discontents.* Cambridge, MA: MIT Press.

Turkle, Sherry. 2010. *Alone Together: Why We Expect More from Technology and Less from Each Other.* New York: Basic Books.

Turner, Fred. 2006. *From Counterculture to Cyberculture: Stewart Brand, the Whole Earth Network, and the Rise of Digital Utopianism.* Chicago: University of Chicago Press.

Turner, Terrence. 2002. Representation, Polyphony, and the Construction of Power in a Kayapo Video. In *Indigenous Movements, Self-Representation, and the State in Latin America,* ed. Kay Warren and Jean Jackson. Austin: University of Texas Press.

Turner, Victor. 1967. *The Forest of Symbols: Aspects of Ndembu Ritual.* Ithaca, NY: Cornell University Press.

Twist, Jo. 2005. UN Debut for $100 Laptop for Poor. *BBC News,* November 17. http:// news.bbc.co.uk/2/hi/technology/4445060.stm.

Uco, Cesar. 2007. Peru's President Garcia Faces Nationwide Protests. Wsws.org, July 20. http://www.wsws.org/articles/2007/jul2007/peru-j20.shtml.

United Nations Conference on Trade and Development—International Centre for Trade and Sustainable Development (UNCTAD-ICTSD) Project on Intellectual Property Rights and Sustainable Development. 2003. *Intellectual Property Rights: Implications for Development.* Geneva, Switzerland: United Nations.

United Nations Conference on Trade and Development (UNCTAD). 2008. Creative Economy Report 2008: The Challenge of Assessing the Creative Economy towards Informed Policy-making. Geneva, Switzerland: United Nations.

UNESCO Institute for Statistics. 2011. Country and Regional Profiles: Peru. http://stats.uis.unesco.org/unesco/TableViewer/document.aspx?ReportId=198.

Unwin, Tim. 2009. *ICT4D: Information and Communication Technology for Development.* Cambridge, UK: Cambridge University Press.

Uy-Tioco, Cecilia. 2003. The Cell Phone and EDSA 2: The Role of a Communication Technology in Ousting a President. Critical Themes in Media Studies Conference, New School University, October 11.

Vaidhyanathan, Siva. 2006. Rewiring the "Nation": The Place of Technology in American Studies. *American Quarterly* 58 (3): 555–567.

Vaidhyanathan, Siva. 2011. *The Googlization of Everything (And Why We Should Worry).* Berkeley: University of California Press.

Verran, Helen. 2007. The Educational Value of Explicit Noncoherence: Software for Helping Aboriginal Children Learn about Place. In *Education and Technology: Critical Perspectives, Possible Futures,* ed. David Kritt and Lucien Winegar. New York: Lexington Books.

Verran, Helen. 2011. Comparison as Participant. *Common Knowledge* 17 (1): 64–72.

Vessuri, Hebe. 2011. La Actual Internacionalizacion de las Ciencias Sociales en la America Latina: Vino Viejo en Barricas Nuevas? In *Estudio Social de la Ciencia y la Tecnologia desde America Latina. 2011,* ed. A. Arellano Hernandez and P. Kreimer. Bogota: Siglo de Hombre Editores.

Vicente, Rafael. 2003. The Cell Phone and the Crowd: Messianic Politics in the Contemporary Philippines. *Public Culture* 15 (3): 399–425.

Vigo, Manuel. 2011. Peru's Economy Expected to Grow in 2012, According to IMF. Peruthisweek.com, December 4. http://www.peruthisweek.com/news-1185-Perus-economy-expected-to-grow-in-2012-according-to-IMF.

Villanueva, Eduardo. 2011. The Importance of Being Local. The European Magazine. April 28. http://www.theeuropean-magazine.com/256-villanueva-masilla-eduardo/255-ict-in-development-cooperation.

Vivieros de Castro, Eduardo. 2004. Exchanging Perspectives: The Transformation of Objects into Subjects in Amerindian Ontologies. *Common Knowledge* 10 (3): 463–484.

Vota, Wayan. 2009. If & When Schools Invest in ICT, Teachers First. EduTech Debate.org. http://edutechdebate.org/ict-in-education/if-when-schools-invest-in-ict-teachers-first.

Wallace, Lewis. 2009. *Wired* Backs Internet for Nobel Peace Prize. Wired.com, November 20. http://www.wired.com/underwire/2009/11/internet-for-peace-nobel.

Wallerstein, Immanuel. 2006. *European Universalism: The Rhetoric of Power*. New York: New Press.

Warren, Kay. 1998. Indigenous Movements as a Challenge to the Unified Social Movement Paradigm of Guatemala. In *Culture of Politics, Politics of Cultures: Re-Visioning Latin American Social Movements*. ed. Sonia Alvarez, Evelina Dagnino, and Arturo Escobar. Boulder, CO: Westview Press.

Warren, Kay, and Jean Jackson, eds. 2002. *Indigenous Movements, Self-Representation, and the State in Latin America*. Austin: University of Texas Press.

Warschauer, Mark, and Morgan Ames. 2010. Can One Laptop per Child Save the World's Poor? *Journal of International Affairs* 64: 1.

Washington Post. 2011. The Netbook Lives! Asus Launches Product Page for the MeeGo Eee PC X101 Netbook. Washingtonpost.com, July 28. http://www.washingtonpost.com/business/technology/the-netbook-lives-asus-launches-product-page-for-the-meego-eee-pc-x101-netbook/2011/07/28/gIQAeGrpeI_story.html.

Weber, Samuel. 1991. *Return to Freud: Jacques Lacan's Dislocation of Pscyhoanalysis*. Cambridge, UK: Cambridge University Press.

Weber, Steven. 2004. *The Success of Open Source*. Cambridge, MA: Harvard University Press.

White, Luise. 2000. *Speaking with Vampires: Rumor and History in Colonial Africa*. Berkeley: University of California Press.

Whyte, William. 2002. *The Organization Man*. Philadelphia: University of Pennsylvania Press.

Williams, Maggie. 2002. Latin America in Love with Linux. vnunet.com, April 23. http://vnunet.com/News/1131173.

Willinsky, John. 2009. *The Access Principle: The Case for Open Access to Research and Scholarship*. Cambridge, MA: MIT Press.

Wilson, Fiona. 2001. In the Name of the State? Schools and Teachers in the Andean Province. In *States of Imagination: Ethnographic Explorations of the Postcolonial State*, ed. Thomas Blom Hansen and Finn Stepputat. Durham, NC: Duke University Press.

Wilson, Fiona. 2007. Transcending Race? Schoolteachers and Political Militancy in Andean Peru, 1970–2000. *Journal of Latin American Studies* 39:719–746.

Winner, Langdon. 1989. *The Whale and the Reactor: A Search for Limits in an Age of High Technology*. Chicago: University of Chicago Press.

Winnicott, David W. 1971. *Playing and Reality*. New York: Routledge.

Wired.com. 2003. Brazil Gives Nod to Open Source. Wired.com, November 16. http://www.wired.com/news/infostructure/0,1377,61257,00.html.

Wong, Tzen, and Graham Dutfield, eds. 2010. *Intellectual Property & Human Development: Currents, Trends, & Future Scenarios*. Cambridge, UK: Cambridge University Press.

World Bank. 2005. Poverty Assessment: Peru. http://web.worldbank.org/WBSITE/ EXTERNAL/COUNTRIES/LACEXT/0,contentMDK:20862551~pagePK:146736 ~piPK:146830~theSitePK:258554,00.html.

World Bank. 2009. *Information and Communications for Development 2009: Extending Reach and Increasing Impact*. Washington, DC: World Bank Publications.

World Bank. 2012. Country Brief: Peru. http://web.worldbank.org/WBSITE/ EXTERNAL/COUNTRIES/LACEXT/PERUEXTN/0,contentMDK:22252133~pagePK :1497618~piPK:217854~theSitePK:343623,00.html.

World Intellectual Property Organization. 2011. Website. Frequently Asked Questions. http://www.wipo.int/about-wipo/en/faq.html.

Yudice, George. 1998. The Globalization of Culture and the New Civil Society. In *Cultures of Politics, Politics of Cultures: Re-Visioning Latin American Social Movements*, ed. Sonia Alvarez, Evelina Dagnino, and Arturo Escobar. Boulder, CO: Westview Press.

Yudice, George. 2003. *The Expediency of Culture: Uses of Culture in the Global Era*. Durham, NC: Duke University Press.

Zaloom, Caitlin. 2010. *Out of the Pits: Traders and Technology from Chicago to London*. Chicago:University of Chicago Press.

Zerda-Sarmiento, Alvaro, and Clemente Forero-Pineda. 2002. Intellectual Property Rights over Ethnic Communities' Knowledge. *International Social Science Journal* 54 (171): 99–114.

Žižek, Slavoj. 2010. *Living in the End of Times*. New York: Verso.

Index

Indigenous peoples. *See also* Cultural
  preservation
  and authenticity, 65–67
  and globalization, 30–31, 64–65
  and IP-based development, 26
  and market forces, 64–65
  and native stagings, 65–67, 72
  poverty rates, xx
  state recognition of, 64
  strategic essentializations of, 65–66
  and unrest, xx–xxi, 3, 177
Indignados, xi, 7
Indonesia, 154
Information Age trilogy, 10
Information and Communication Technologies (ICTs)
  and cultural traditions, 14–15
  and diverse participation, 12–13
  export partnerships, 15, 47, 97
  and India, 197
  ICT4D projects, 179
  and logics of inclusion, 13
  and peripheral actors, 13–14, 19
  and universalism, 19, 180
Information culture. *See* Digital culture
Information economy, 8, 26, 30–31, 66,
  148
Information flow, 4
Information society, xiii, xviii, 9–11, 19
Inga, Max, 56
Innovation, technological, 132. *See also*
  Technological universalism
Innovation centers, ix–xi
  and Internet diffusion, xii
  and popular movements, 6–8
Innovation classrooms, xii–xiii, 11
Innovation language, 177–178, 203
Intel, 198
Intellectual property, 1–2, 14, 125. *See
  also* Denomination of origin (DO);
  IP titles
International Labor Organization's Con-
  vention No. 169, 63
International Monetary Fund (IMF), xiv

Internet
  and democracy, xii, 6–7
  global spread of, 5
  and innovation centers, xii
  and Nobel Peace Prize, 6–7
  and social movements, 6–8
  as "tool for peace," 7, 17
  and traditional politics, 17
  and universalism, 18–19
  and utopianism, 17, 186, 189
Internet cafes, 161
Interrupt, digital, 176, 194–195
IP titles, 1–2, 8, 199–200. *See also* Chulucanas ceramics; Denomination of
  origin (DO)
  and authorship, 204–205
  and competition, 205
  and culture, 33
  exclusionary function of, 33–34
  global expansion of, 25–27, 33
  government policy, 32–33, 50–51
  and market potential, 33
  and native products, 27–28
  and optimization, 33, 35, 61–62
  and rural development, 11, 15, 23–24,
  27, 32–33
Iran, 6
Irani, L, 5
Isaac, G., 26

Jackson, J., 64
James, Jeffrey, 198
Jaramillo, I. C., 160
Jarvis, J., 5, 6, 156
Jauregui, C., 98
Jenkins, H., 119
Jobs, Steve, 162
Johnson, S., 6
Joseph, G., 42, 66
Joyce, Gloria, 77

Kant, Immanuel, 18
Kapczynski, A., 10, 11, 25
Katz, E., 119, 197

Printed in the United States
by Baker & Taylor Publisher Services